# DYNAMIC CONSULTATION IN A CHANGING WORKPLACE

# Dynamic Consultation in a Changing Workplace

*edited by*

EDWARD B. KLEIN, Ph.D.
FAITH GABELNICK, Ph.D.
AND PETER HERR, M.A.

PSYCHOSOCIAL PRESS
MADISON, CONNECTICUT

Copyright © 2000, Psychosocial Press

PSYCHOSOCIAL PRESS ® and PSP (& design) ® are registered trademarks of International Universities Press, Inc.

All rights reserved. No part of this book may be reproduced by any means, nor translated into a machine language, without the written permission of the publisher.

**Library of Congress Cataloging-in-Publication Data**

Dynamic consultation in a changing workplace / edited by Edward B. Klein, Faith Gabelnick, and Peter Herr.
    p.  cm.
  Includes bibliographical references and index.
  ISBN 1-8878-41-26-1
  1. Organizational change. 2. Consultants. 3. Organizational effectiveness.
I. Klein, Edward B.  II. Gabelnick, F. G. (Faith G.)  III. Herr, Peter.

HD58.8.D94 1999
658.4'063—dc21

                                                  99-048851

Manufactured in the United States of America

*This book is dedicated to our children
Laura Ellen and Karen Lynn Klein
Deborah Anne and Tamar Miriam Gabelnick
William, Lucas and Kristine Herr*

# Contents

Contributors ix
Introduction
    *Edward B. Klein, Faith Gabelnick, and Peter Herr* xiii

**Part I: Organizational Consultation**

1. Consultation and Transformation: Between Shared Management and Generative Leadership
   *David Gutmann and Ronan Pierre*     3
2. The Changing Psychological Contract in the Workplace
   *Boris M. Astrachan and Joseph H. Astrachan*     33
3. The Tiller of Authority in a Sea of Diversity: Empowerment, Disempowerment, and the Politics of Identity
   *Geoffrey M. Reed and Debra A. Noumair*     51
4. A Salt-and-Pepper Consultation: Racial Considerations
   *Beverly L. Malone*     81
5. Coping with a Divestiture: The Psychological and Managerial Dilemmas of Ending Personal, Work, Organizational, and Community Relationships
   *Laurence J. Gould*     97

6. The Consultant's Use of the Inner Dialogue in Family Business Consultation
   *Joseph Rosenthal and Donald A. Davidoff* — 113
7. Understanding the Impact of the Founder's Legacy in Current Organizational Behavior
   *Alexander H. Smith* — 133
8. In the Presence of the Other: Developing Working Relations for Organizational Learning
   *Susan Long, John Newton, and James Dalgleish* — 161
9. The Consultant as Container
   *Edward B. Klein* — 193

**Part II: Consultation in Health Settings**

10. Downsizing and the Accidental Consultant
    *Peter Herr* — 211
11. AIDS and the Organization: A Consultant's View of the Coming Plague
    *Burkard Sievers* — 233
12. Training Group Therapists for the 21st Century
    *Mary Nicholas and Robert H. Klein* — 251
13. Changes in the Professional Marketplace: A Personal Odyssey
    *William Hausman* — 275

Afterword
   *Faith Gabelnick, Edward B. Klein, and Peter Herr* — 295

Name Index — 301
Subject Index — 305

# Contributors

**Boris M. Astrachan, M.D.,** is Distinguished Professor of Psychiatry, University of Illinois at Chicago.

**Joseph H. Astrachan, Ph.D.,** is the Wachovia Chair of Family Business at the Michael J. Coles College of Business of Kennesaw State University, Kennesaw, Georgia.

**James Dalgleish** is an organizational consultant in Australia.

**Donald A. Davidoff, Ph.D.,** is an Associate of the Levinson Institute, Waltham, Massachusetts, McLean Hospital, Belmont, Massachusetts, and Harvard Medical School, Boston.

**Faith Gabelnick, Ph.D.,** is President, Pacific University, Forest Grove, Oregon.

**Lawrence Gould, Ph.D.,** is Professor of Psychology, The City University of New York, and Director, Program in Organizational Development and Consultation, The William Alanson White Institute, New York.

**David Gutmann, Ph.D.,** is Executive Chairman, Praxis International, Executive Vice President, International Forum for Social Innovation (FIIS-IFSI), and Professor, Paris Institute for Political Studies, Paris, France.

**William Hausman, M.D.,** is a Professor Emeritus of Psychiatry, University of Minnesota, and an Associate of The Levinson Institute, Boston, Massachusetts.

**Peter Herr, M.A.,** is a graduate student, Psychology Department, University of Cincinnati, Cincinnati, Ohio.

**Edward B. Klein, Ph.D.,** is Professor of Psychology and Psychiatry, University of Cincinnati, and Faculty, Cincinnati Psychoanalytic Institute, Cincinnati, and Board Member, A. K. Rice Institute, Jupiter, Florida.

**Robert H. Klein, Ph.D.,** is Clinical Associate Professor of Psychiatry, Yale University School of Medicine, New Haven, Connecticut.

**Susan Long, Ph.D.,** is Professor of Organisational Dynamics and Director of Graduate Programs in Organisation Dynamics in the Graduate School of Management at Swinburne University, Melbourne, Australia.

**Beverly L. Malone, Ph.D., R.N., F.A.A.N.,** is President, American Nurses Association, Washington, D.C.

**John Newton** is Principal Lecturer in Organisation Dynamics and Program Manager for the Master of Business (Org. Dyn.) in the Graduate School of Management at Swinburne University, Melbourne, Australia.

**Mary Nicholas, Ph.D.,** is in private practice in New Haven, and adjunct faculty, Department of Psychiatry, Yale University School of Medicine, New Haven, Connecticut.

**Debra A. Noumair, Ed.D.,** is Associate Professor, Department of Social, Organizational, and Counseling Psychology, Teachers College, Columbia University, New York.

**Ronan Pierre** is Military Selectee, French Navy, and a member, in charge of the "Youth Forum" Project, of the International Forum for Social Innovation, Paris, France.

**Geoffrey M. Reed, Ph.D.,** is Assistant Executive Director for Professional Development, Practice Directorate, American Psychological Association, Washington, D.C.

**Joseph Rosenthal, Ed.D.,** is on the staff of Andersen Consulting, Wellesley, Massachusetts.

**Burkard Sievers, Dr.,** is Professor of Organization Development/ International Management in the Department of Economics, Bergische University, Wuppertal, Germany.

**Alexander H. Smith, Ed.D., A.B.P.P.,** is in private consultation practice, Cincinnati, Ohio.

# Introduction

*Edward B. Klein, Ph.D., Faith Gabelnick, Ph.D., and Peter Herr, M.A.*

We live in a dynamic world where just standing still means falling behind. The global growth and acceptance of organizational consulting stems in part from the recognition of institutional and environmental complexity. Many leaders struggle to find their place and voice in this rapidly changing world by embracing the support and guidance offered by organizational consultants.

The contributors to this volume, from Australia, Germany, France, and the United States, have consulted to, or worked in, various organizations and recognize the need to connect the internal psychological landscapes of executives, managers, and employees to their external manifestations. Here they present rich case material in order to provide an understanding of how consultants work to facilitate organizational transformation.

The complexity of the cases presented mirrors the complexity that consultants face and will continue to face in this rapidly changing world. The material offers a rich opportunity for leaders, consultants, and consultants-in-training in all fields. In particular, managers, hospital administrators, human resource personnel, executives, consultants, and students and professors of organizational consultation will find the cases illustrative of the dynamic workplace issues they will encounter.

We (Klein, Gabelnick, & Herr, 1998) have reported on the impact of massive sociotechnical and psychological changes in organizational leadership. In this environment, leaders and those they employ to assist in facilitating change have to operate using different models for leading and different approaches to the workforce. Changes are occurring as these boundaries become more fluid, as person-role boundaries blur, and as new opportunities and challenges occur in the external environment. One dramatic result of these transformations is the recognition that the psychological and practical work contract has changed. Leaders can no longer count on the uncritical support of their employees, and employees cannot any longer count on lifelong employment in one organization.

This end to the traditional psychological contract of mutual loyalty and rewards for that loyalty makes many employees feel like temporary hires and makes employers feel less secure in planning for the future. The rules of the game are changing: The old patriarchy with its embedded series of checks and balances and its patrician, benign despotism is yielding to a less centralized, less hierarchical but also less reliable set of organizational relationships. Navigating in this world is like being in "whitewater" all the time. The norm is change, but the yearning is for calm waters and a sense of security and predictability.

Executives are trying to cope with and manage this change, but they also realize that they need assistance. Ironically, the role of consultant with its flexible role definition, its ability to locate at many different sites, and its service orientation mirrors precisely the emerging structures in the workplace. Thus, consultants are being used extensively by progressive companies to assist administrators and employees in understanding change and managing it strategically.

As an outsider, the consultant offers a perspective that is neither "loyal" nor "disloyal," but, rather, neutral and systems-oriented. Effective consultation in these circumstances, therefore, focuses not only on operational change but also on the psychodynamic culture in which the change is to occur.

Any change that can be effected will endure only if that change is encased in a reflective, open system. The task of the consultant as he or she develops a working relationship with management is to emphasize an open-ended facilitative style of leadership and to communicate a respect for the difficulty of learning to change and manage the change in a developmental framework.

In this current and near-future work environment, consultants and faculty who train future organizational consultants will have to use approaches that enable this type of workplace to grow in responsive and responsible ways. Models of leadership are already changing, and institutions are reflecting the experience of working with these new approaches. However, what is less developed is a *view* of leadership and consultation that empowers and creates transformational thinking and living.

We assert that attempts at transforming organizations often falter because of unrecognized or unacknowledged anxiety (and resistance) to change. Consultations will require a more sophisticated approach to institutions, one that includes integrated recognition of the impact that different cultures, ethnic groups, genders, and even languages can have on how individuals work together. Thus, managers and consultants who function in workplaces today and in the future will need a dynamic systems perspective or theoretical framework in order to link proactively the people and the organization.

Similarly, consultants are now working in new ways. A subtext of this volume is that the act of consultation itself will be managed successfully only if the consultants are continuously monitoring internally the impact that these issues have on their own lives and work. The interrelated, intimate nature of the consultants' work with the institution makes it important for consultants to manage their own psychodynamic landscape and in turn seek assistance in maintaining a balanced perspective on their work.

The psychological and systems perspectives used in this volume assist readers in developing their own ideas about work, role, task, and person. It is a contemporary book in that its thrust is to enable and facilitate reflection, not just to provide recipes for organizational improvement.

The volume is divided into two sections: Organizational Consultation and Consultation in Health Settings. In Part I, "Organizational Consultation," general issues affecting modern institutions are spelled out in nine chapters.

In Chapter 1 David Gutmann and Ronan Pierre, organizational consultants based in Paris, France, discuss consultation and transformation between shared management and generative leadership, using European and American political, economic, and corporate examples. Their consulting work led to an institutional transformation approach to organizations, taking into account the unconscious in its institutional expression. They focus on the leader's ability to comprehend a complex environment, resistance to change, and how to transform past activities into future innovations.

In Chapter 2, Boris Astrachan and Joseph Astrachan discuss the changing, unspoken psychological contract in the workplace. With the end of long-term employment and huge pay differences between top management and workers, organizations cannot count on the whole-hearted support of employees. Thus, major industrial and health institutions need to find ways to develop job security and career protection and to reduce the focus on excessively compensated executives in order to improve work relations and productivity.

In Chapter 3, Geoffrey Reed and Debra Noumair describe their diversity training activities in a university setting, where individuals often identify themselves as victims. They address the organizational dynamics of various groups fighting for resources by questioning the overall goals and institutional authority of diversity work. Given the ethnic differences found in the modern university and institutions in general, these dynamics will be even stronger in the future (witness the passage of the affirmative action proposition in California and the prediction that Whites will be a minority in the United States by the year 2050).

In Chapter 4, Beverly Malone describes her personal and professional experiences as part of a forced salt-and-pepper consulting team. The multiracial female pair were used to mask the unresolved issues of a large southern health care organization. The very composition of the consultation team was prognostic of the failed Black and White attempt to save the organizational leader who was the author's friend. The collusive lack of perspective and denial of impending danger by the client and the consultants limited the effectiveness of their work. As the Black consultant of a less than fully functioning team, Dr. Malone also contained much of the racial dynamics in the institution and community.

In Chapter 5, Laurence Gould sensitively describes the anxieties and dilemmas involved in ending interpersonal and work relationships while managing a divestiture of an oldtime energy company. With the increasing number of mergers, acquisitions, and takeovers in industry and health care, executives and external consultants need to more effectively address the conflicts and tensions caused by terminating personal, community, and work relationships while effectively managing a divestiture. This phenomenon is paramount in the industrial world now, and consultants will continue to work with such issues for years to come.

In Chapter 6, Joseph Rosenthal and Donald Davidoff provide clear examples of how the use of personal history and self-awareness helps consultants work competently with a family business. Since the majority of Fortune 500 companies began as family businesses, we need more examples of how the experience and insight of external consultants helps them work effectively with executives in family businesses. That is, the consultants' own family histories and their reactions to the dynamics of the family business enhance their understanding and the effectiveness of their work. Consultants need to share such case studies with social scientists and managers to help produce more effective coping strategies in the future.

In Chapter 7, Alexander Smith presents a number of issues critical to consulting in different Catholic religious orders. The consultant has to be extremely sensitive and clear in very demanding, historically determined situations. The founder and his or her vision have great influence on present members of the community. There are many parallels here to business organizations with charismatic founders; both attract with their unique vision and require consultants to appreciate these visions in order to be effective.

In Chapter 8, Susan Long, John Newton, and James Dalgleish of Swinburne University in Melbourne, Australia, present a collaborative action research project with a changing government organization. The authors discuss some similarities and differences between research and organizational consultation. They also explore some fascinating parallels (pressures to be more economically competitive and being left out of decisions) between the university and the government agency that were explored in the action research/consultative process. Both groups were enriched by the $1^1/_2$-year collaborative process of making the mandated changes more rational and less hurtful.

In Chapter 9, Edward Klein notes that given the stresses relating to changes in organizational boundaries, an increasing problem involved in long-term consultation is countertransference or the projecting of staff's emotions and conflicts onto the consultant. This phenomenon occurs as the consultant becomes the vehicle for the expression of employee anger or frustrations. One way to effectively address this problem is for the consultant to work with a senior "shadow" consultant. This is an approach similar to that of a clinician using a very experienced outside supervisor in order to obtain a more objective perspective on the dynamic processes involved in the therapist-patient relationship.

Part II, "Consultation in Health Settings," includes four chapters. The health care industry currently represents one seventh of the economy and is undergoing drastic changes as organizations deal with intense competition, rapidly advancing technology, highly educated professionals, and some of the most difficult social issues of our time. Health care is also an excellent example because these institutions have been open to innovation and experimentation and thereby provide models for future leadership and consultation.

In Chapter 10, Peter Herr describes an accidental consultation he provided as the health care organization he worked for underwent radical transformation and downsizing. As a doctoral student he was increasingly trusted by fellow employees and provided a sensitive ear and critical systems viewpoint that helped colleagues understand and adjust to dramatic changes in the institution. This chapter is an example of what happens when a noninclusive management style fails to solicit and integrate employee contributions.

In Chapter 11, Burkard Sievers describes his consultation to German AIDS treatment organizations whose basic assumptions changed when their personnel were confronted with divisions based on which staff faced imminent death. Sievers offers a way of understanding and working with this plague of our time, which continues to divide people based on sexual preference and infection status.

In Chapter 12, Mary Nicholas and Robert Klein note the training necessary to be an effective group leader while dealing with the economic constraints of current health care systems. They describe the compromises that consultants make in order to do the training in cost-containing mental health care organizations that are less supportive of educational activities. Their work in various clinical and educational institutions highlights some of the intellectual, developmental, and emotional costs—for both staff and patients—of managed care now and in the future.

In Chapter 13, William Hausman shares his experience of being a *locum tenens* and the problems of a temporary leader, which are major issues in our technological, economically driven society. He notes that professionals, like doctors, are now and will be in the future, temporary and temporal—hired for the skills they can deliver on a particular project. Institutional loyalty becomes a quaint phrase, while time spent on task and productivity are the guidelines for "success" and survival. Even with these depersonalized employment patterns, he notes important variations in how these forces are played out and expressed in different health care settings.

We support a more internally complex, integrated consultant stance in the face of external chaos and internal organizational instability. The strength and flexibility of administrators and consultants are tested as they strive to handle organizational ambiguity. This type of situation represents a very challenging task for both the consultant and the manager. As the contributors to this volume demonstrate, consultants are transformed by working in organizations and thus need to educate leaders on more collaborative management styles despite disconnected employment patterns.

There is little doubt that the transformations discussed in this volume create anxiety for both leaders and those they lead. How external consultants help executives manage these anxieties will determine managerial success in the workplace. It remains unclear whether the work team will be a satisfactory container of employee emotion. While a call for workers to creatively link together is an ideal, paradoxically, connectivity runs counter to individualism and self-reliance. New visions will have to contend with the old myth of the heroic leader/manager.

These complex current and future developments suggest that it is a combination of psychodynamic and open systems theories that most effectively helps managers and consultants understand the nature, quality, and structure of institutions and the roles of individuals within the organization. It is our hope that such an understanding will translate into more competent management and organizational consultation as we move into the 21st century.

## REFERENCE

Klein, E. B., Gabelnick, F., & Herr, P. (1998). *Psychodynamics of leadership*. Madison, CT: Psychosocial Press.

# Part I

## Organizational Consultation

# 1

# Consultation and Transformation:

## Between Shared Management and Generative Leadership

*David Gutmann, Ph.D., and Ronan Pierre*

The cloning process achieved by Scottish scientists with a sheep called Dolly shook the whole world in March 1997. The date is likely to be remembered by future generations: They will definitely have to confront the possibility of human cloning, despite all the reassuring speeches about prudence in experimenting. In this context, it is not difficult to relate to our science fiction literature background. From H. G. Wells to Philip K. Dick, which author has not written a story of human cloning? Traditionally, the human cloning theme is linked to two other themes in science fiction: immortality and/or totalitarianism. Cloning is a way to be immortal by copying oneself as well as to produce masses of identical individuals, a basis for obedient and interchangeable servants. In these extreme representations of society—themes that are reflected in science fiction—we observe two tendencies of *leadership* which unfortunately occur too frequently: Leaders tend to monopolize power and leadership.

In this chapter we shall examine the relationship of leadership and transformation, and the importance of consultation. We consider that the first task of the consultant (a role we call "adviser in leadership") *is helping the leader to avoid these temptations of becoming all-powerful and all-knowing.* Our assumption is that leadership cannot be analyzed by itself, but finds meaning in the internal and external environment in which it takes place. The role of the consultant is helping to detect the conscious and, even more, the unconscious processes in leadership as it manifests itself in the external environment and in the psyche of the leader.

Our approach is centered on an *Institutional Transformation (IT)* model of organizations, which takes into account the unconscious in its institutional expression. By institution, we mean any social organization or structure with a primary task, like a firm, an association, or a football team. By *transformation,* we mean *passage from one form (-ation) to another,* the process of which we consider to be a condition for innovation. Our experience of transformation is directly related to our work with leaders in helping them to comprehend their transient, complex environment. As consultants, we try to see what resistances are at work in organizations when a leader tries to transform and "govern" an institution. We have to remember at this point the very enlightening etymology of *"leadership."* "Leader" derives from the Germanic word *leiten,* which refers both to the path and the convoy: A leader is the one who conducts the convoy and/or shows the path. *Ship* relates also to the Germanic root *skap,* which means to create, or to shape. From this perspective, leadership means to shape one's way, to shape the path and the convoy together.

The thesis we are proposing here is that no institution can sustain itself without transforming itself (Part I). A leader must take the helm of the transformation process in order to guarantee its success, but there is no leadership without transformation (Part II). This implies that leadership has to be generous and generative (Part III). Therefore, as advisers in leadership, we must focus our intervention on institutional transformation. The consultant's role within the institution then completes a *sustaining triangle* composed of the leader, the manager, and the consultant, which is essential to the institution's dynamics (Part IV).

## I. NO INSTITUTION CAN SUSTAIN ITSELF WITHOUT TRANSFORMING ITSELF

A perennial institution is the dream of any manager. This is especially the case in the economic sector, where firms of any size are organizations that, besides their collective primary task (generally profit), provide a livelihood for their employees. However, a perennial institution also takes the risks of being outmatched by other organizations in its environment with the same primary task. This principle is at the core of the classical economic theory of Adam Smith and Jean Baptiste Say that gave birth to our current capitalistic system. Later, Schumpeter (1939) analyzed the evolution of firms in a competitive environment as a "creative destruction" process: When a firm enters the market with a new product its success is short, for the innovation is somehow integrated by its environment and new products appear, cheaper or of better quality. Therefore, the firm is confronted by innovative new tasks. Often, the firm disappears. This destruction, however, has been a gain for society, since it has brought new and cheaper products. Therefore, it is a "creative destruction" process.

This economic scheme has a counterbalance in the psychoanalytical field. We perceive that an institution, like an individual, is constantly confronted by a question of self-renewal, and can be involved in a process of *repetition*. In keeping with this Freudian concept (Freud, 1920/1989), we recognize the tendency of the subject to reproduce sequences that originally created suffering. The impossibility of obtaining the desired thing creates suffering (repetition is guided by the pleasure principle in the first place, but its obvious relationship to suffering and pain induced Freud to introduce the notion of "beyond the pleasure principle"). In the same manner, we believe that repetition can sometimes place the institution in danger. In this case, the danger is either concrete and physical (destruction, as in the Schumpeter theory), or more subtle, diffusing itself within the institution and endangering members (again, either physically or psychically). Additionally, despite the pain it creates, repetition serves as a defense from anxiety (Gutmann, 1988), because fear of the unknown generates high levels of anxiety. We can say that for an individual or an institution, attempting to avoid repetition is often much more difficult than the search for stability, which brings less anxiety within the institution or for the individual.

If stability is threatening, could instability, the opposite state, be less endangering? Of course not. But it is a condition for innovation, for avoiding repetition. Any effective innovation demands time and energy, and often entails much more anxiety than leaving things "untouched," "untransformed." In our experience, innovation can only be achieved if an institutional transformation process is at work. What is transformation? As consultants, how can we detect and aid transformation?

Unfortunately, the "modern" conception of management and leadership constitutes a science that promises managers and leaders that they will be able to analyze, forecast, and monitor behaviors in the organizations for which they are responsible. How can this be so? In the history of science, we are only beginning to comprehend the complexity of human behavior (Foucault, 1966).[1] Freud made his major contribution to the discovery of human nature only a century ago. In discovering the unconscious and the oedipal determinism of human being, our civilization also became aware of the frailty of the myth of human self-possession. With regards to this history of science, one should be careful and modest when claiming that management is a science that can be taught or learned.

---

[1] "The danger [of this enterprise], in short, is that instead of providing a basis for what already exists . . . , one is forced to advance beyond familiar territory, far from the certainties to which one is accustomed, towards an as yet uncharted land and foreseeable conclusion. . . . Is there not a danger that everything that has so far protected the historian [consultant, manager, professional . . . ] in his daily journey and accompanied him until nightfall . . . may disappear, leaving for analysis a blank, indifferent space, lacking in both interiority and promise?" (Foucault, 1969).

The other main reason why forecasting and monitoring human behavior is a fantasy is that the role of representations, imagination, affects, and drives is, of course, determining in the life of an institution. On this particular point, it is interesting to note that a growing number of scientists, in domains in which a strong rationale and "Cartesian" culture traditionally prevails, uses some intellectual material or even corroborates some results of the social sciences. For instance, the role of emotions in the decision process of an individual is now a hypothesis in which neurologists take an interest. In order to confront all the choices that an individual can make, the frontal cortex of the mind elaborates images of the different scenarios of actions to be taken. These images create an emotional reaction linked to their content. Emotions are generated by a "change of the corporeal landscape"; each emotion connects with a different setting of the body. This theory of the role of emotions alleviates the belief in the rationality of the decision process (Damasio, 1995, p. 183). Though its reality is not established, this theory enables us to understand the complexity of decision making, taking into account some "irrational" material in management, for example. Thus, consultancy inside organizations has to take into account these aspects. This is why we consider the dynamics of organizations in terms of "institutional transformation."

Our point of view is that the transformation process is not simple. Each institution is an open system to the transformation forces that come either from the interior or the exterior. The institution integrates these forces but, according to the *homeostasy*[2] principle, works it out; it resists and then modifies these forces in order to keep itself stable as long as possible, to keep the system as it is. For this, several modes of resistance are used.

Often, when a director or an executive, a ruler or a manager, formulates consciously the future of his institution from state A, he represents it by an objective; let us call it state B. Contrary to

---

[2] The principle of "homeostasy" was largely used by cyberneticians to refer to the capacity of an organization to maintain its equilibrium. According to this principle, born along with the medical progress of the late 19th century, all living mechanisms, as diverse as they can be, have only one goal: to keep unity of their living conditions within their environment.

what most "managerial" experts claim, *there is no evolution of the institution that can be rationalized or forecast*— there is no "from A to B" (a "To Be"). *B is a fantasy. The closer you get, the more it transforms itself.*

We can explain this phenomenon with the help of chaos theory (Dahan Dalmedico, Chabert, & Chemia, 1992). Since the institution is a product of human behavior and imagination, it is chaotic in the sense that no rational prediction can be made about its evolution. According to chaos theory, in terms of evolution, a small disturbance in initial conditions can result in huge disruptions after a time. An obvious example is the weather. A famous saying is the flap of a butterfly's wings on the banks of the Chiang Jiang River in China can generate a hurricane in the Atlantic Ocean. Chaos, however, is not disorder, for chaotic states are determined—but unpredictable. In a similar way, the future of the institution cannot be predicted, because of all the different disturbances (external and internal) that can affect its evolution, but it is determined, which means that it can only evolve toward determined states.

The passage from one state to another *(trans-formation)* is neither regular nor continuous: It is chaotic, erratic, and discontinuous. Because it comes up against irregular resistance, the movement of transformation adopts a path composed of *zigzags,* progression and regression, construction and destruction. Let us recall our own experience: A personal or an academic failure can also be a way to spring back to life again. We even think that *the presence of these resistances are clues confirming the authenticity of a process of transformation,* distinguishing it from a simple measure of superficial regulation.

To schematize, this sinuous passage is comprised of blurred phases that are strange if not incomprehensible, and seem lost in the obscurity of a "black box." By definition, these segments of the passage are never seen, known, or understood. The black box is the uncontrollable, determined, but unpredictable part in the evolution of the institution. Inside the black box one could find all the "nonrational" elements present in the interactions between members of the institution and between the institution and its environment, in which affects and unconscious drives are determining.

Other segments, as if they extend beyond the black box, are identified and apprehended by certain members of the group. For example, a rise in the turnover rate within a firm can be an indicator of future difficulties. These points emerge like tips of an iceberg above water: They necessarily imply the existence of a hidden mass that is even larger, beyond what is apparent. In other words, it is a question of an unknown byway that is long and tangled, lost in the darkness of the black box. In addition, these visible segments do not, alone, allow the plotting of the totality of the path, or even draw a view of the whole.

Other steps along this path are visible but untouchable; they seem out of reach or outside the capacity for action, influence, and regulation by the members of the group. Often, they seem unreachable because it seems that transforming them would entail consequences too dangerous for the institution. These elements can be found in a transparent "glass box" that itself contains the black box (see Figure 1).

We think that *if the leader incites people to believe that B is an absolute certainty (instead of a fantasy) it will be demoralizing* for

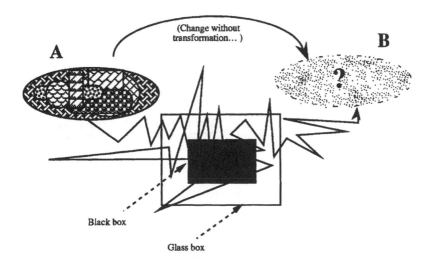

Figure 1. The path of transformation.

those concerned with the primary task. This is especially the case in Western society. In contrast to our civilization, the Chinese culture, for example, both in its Taoist and Buddhist aspects, acknowledges that one can never see the whole of a thing (Jullien, 1992), which fosters a certain relativism in considering life (Figure 2). In a sense, accepting the unknown is also accepting the Other.

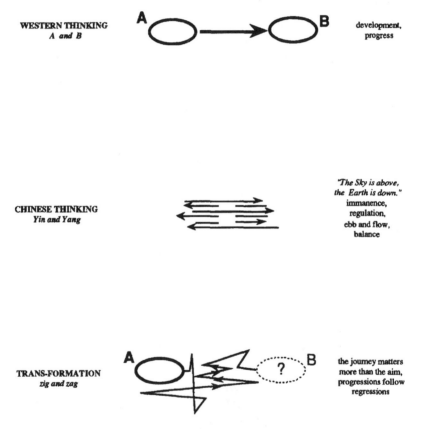

Figure 2. considering trans-formation.

Each one of us, in our personal and professional life, goes through a transformation process, indeed, *is* in transformation. Thus, he or she lives periods of progression and regression. Is it possible to deal with these moments of regression, to accept them, when one is at the head of a corporate institution? In our work, we often observed that decision makers who had had a model career (top universities, important jobs, successful achievements) were particularly vulnerable to a sudden obstacle or misfortune in their lives. In that case, just like a revelation process during a psychoanalytical cure, they went through depression, but the problem was that such individuals, being in situations of power in their institutions, were also harmful for these institutions at these times in their lives. They often became omnipotent or, to the contrary, invisible in their exercise of leadership. As Freud (1917/1989) observed, the depressive structure (melancholy) is characterized by the impossibility of the person mourning the lost object. This particular task is at the core of leadership "ethics": The leader is responsible for taking into account his or her full personality, in fulfilling his or her role in the institution. The consultant can help in this work.

Just like individuals, institutions are confronted by the incapacity or unwillingness to accept regression, which can make it either vulnerable to brutal inner or exterior evolutions, or omnipotent. For example, Microsoft holds 80% of its market, and has been growing since 1983. It appears to have mastered every difficulty. But what can happen to a firm that is not in the habit of confronting regression, in case of unforeseen trouble—for example, the huge success of the Internet? Will it be able to transform itself? This question can also be raised concerning leaders or managers.

First, in order to help institutional transformation, the consultant in the role of adviser in leadership must use the information in the glass box. Some elements of the institution can be revealed and worked out, which encourages innovation in individual and collective behavior. Working with elements that are observable in the glass box is a *heuristic*, process (discovering facts). It is also the consultant's role to reduce the unknown, the black box, whenever possible. At this step, the consultant can only interpret. The black box is unobservable. For this task the consultant engages in a process of *hermeneutics* (the science of interpreting facts). Thus, being an adviser includes both heurism and hermeneutics. However, the heuristic process must be at the core of a consultant's practice.

Often this process—either detecting facts or formulating working hypotheses about what is happening inside the black box—is disturbing for the members of the institution. The consultant points out certain elements that were "forgotten," cast aside. We consider that conflictual matters have to be worked out and not systematically avoided. To make them discussable is a way of regulating and containing them. Conflicts are therefore reframed constructively.

The consultant's role is to disturb; he or she says things that can hurt. But this approach is necessary so that distressing issues can be discussed and worked out in the "here and now." To formulate publicly a working hypothesis is a way to give birth to a debate and to generate encounters. In the most productive cases, such a practice can reveal habits, prejudices, defensive attitudes, fears, fantasies, anxiety—all elements that can inhibit the individual in his or her relationship with others, and which are resistance in the social innovation and transformation processes. Progressively, by a series of working hypotheses, we can hope that some "hidden elements"[3] of the institution will be revealed.

There are some elements that make transformation especially difficult or even impossible. These are often related to the notion of "difference" or to the incapacity of members of an institution to accept cooperation. The words "boundaries," "borders," "barriers," and "barricades" form a progression in terms of increasing demarcation. For transformation, it is necessary that "barricades," which are impermeable, become "barriers," then "borders," then "boundaries," which can be progressively more permeable. Institutional transformation might involve moving from a closed conception of the limit ("the barricade") to a more open conception ("the boundary"). It expresses the idea that a boundary marks both a rupture *and* a passage; without boundary, there is no differentiation, but also no transaction.

---

[3] René Girard (1978), in *Des choses cachées depuis la fondation du monde* (literally, *About Things Hidden Since the Origins of the World*), defended the idea that the "scapegoat" phenomenon is constant and universal in human groups. In our sense, the "hidden elements" are the constant (anthropological) elements of all institutions. Most are hidden to those who don't want to see them. Some will always stay hidden except when a crisis allows us to open the black box.

This idea of *passage* is also important in the leadership phenomenon. Because leadership has to be both generous and generative, a leader must generate other leaders and accept the fact that they deviate from the path he or she has chosen. In a succession process, for example, both a rupture and a passage must be achieved.

## II. NO TRANSFORMATION WITHOUT LEADERSHIP AND NO LEADERSHIP WITHOUT TRANSFORMATION

What are the links between leaders and the dynamic structure that supports them? Historical examples of leaders' assassination as an expression of unconscious global resistance are numerous: Abraham Lincoln, John F. Kennedy, Robert F. Kennedy, Martin Luther King, Indira Gandhi, Anwar al-Sadat, and recently, Yitzhak Rabin. All assassinations were linked to global political issues and were ultimately expressions of resistance incurred by the leader. It seems as though whenever a leader emerges, unexplained forces, which express themselves through a defined element (in this case, a murderer), would depose or demolish this leader. Overall, leadership has to be understood as a complete and global phenomenon. Transformation and leadership are related in a complementary way.

First, our point of view is *that there is no transformation without a leader*. This necessity is so strong that talking about a "transformative leader" is, in a sense, redundant. This statement can be interpreted as a general paradox of democracy. Although everyone is equal, no task can be achieved by itself, and someone to lead is desperately needed. Thus, each individual might be called upon to take on a leadership role at least once in his or her lifetime.

Second, without transformation, the institution is both threatened and threatening in its interactions with the leader. *No leadership can exist without transformation* (of its role and of the institution). For example, two kinds of altered leadership can occur if transformation is lacking. In both cases, the leader is confronted by the risk of "psychic exhaustion."

A first possibility is that a *superleader* can arise. In this case the leader is flattered by the surrounding entourage. Positive projections are all he or she receives: trust, congratulations, success. The leader cannot play a role of protecting the group, or guarding the institution's boundaries; thus, both the leader and the group are exposed to destructive projections. The suicide of Kurt Cobain, lead singer of the rock group Nirvana, is a good example of such a phenomenon. In this case, a quick and huge worldwide success became in fact . . . Nirvana's Hell. Before Cobain, famous rock or movie stars also went through a self-destruction process: Jimi Hendrix, Janis Joplin, Jim Morrison, Brian Jones, and Marilyn Monroe.

Alternatively, the leader can play the role of a *scapegoat:* This happens when the leader receives negative projections such as envy, destruction, suspicion, rivalry. The scapegoat, or, more precisely, the sacrificial process is not wanted consciously by members of the institution, but is functional: It restores order in the organization by allowing internal conflicts to focus on a victim (Girard, 1972). These two "distortions" of leadership are typical of nontransformation states.

Besides these two types of altered leadership, the nonexistence of a transformation process can produce, out of institutional transformation, an autocratic leader, if not a dictator. If this happens, the institution engages in a vicious cycle, which we call the *Savior, Victim, Persecutor Circle*. This happens when the leader emerges in a particularly difficult context—for example, a well-known manager hired by shareholders in order to help a firm face its great losses or threatening competition, or a young and therefore promising politician elected in a crisis context. He or she is first the institution's Savior. A politician or a company executive taking up his or her post stirs up tremendous hope . . . and actually may lose popularity soon afterward. This phenomenon seems inherent to all leadership takeovers in an institution in crisis. Confronted by important resistance, the leader is then an easy prey for critics, and rapidly becomes a scapegoat for all the aches of the institution, receiving strong negative projections. The leader thus becomes the Victim. In reaction, the leader tends to concentrate power in his or her hands, makes a series of hazardous and arbitrary decisions, and thus becomes the Persecutor.

Overall, the emergence of a leader is a complete, global, and systemic phenomenon. The consultant can only help to clarify the emergence of unconscious resistance that alters leadership—but not the emergence of leaders.

During the 20th century, we can say that some transformation processes have actually occurred. For example, we see several transformation "peaks" in the United States, represented by Franklin Delano Roosevelt's presidency (the New Deal, the decision to enter World War II) or John F. Kennedy's mandate (Medicare and Medicaid)—the concrete results of which still prevail today. Since 1973, the end of the economic "Golden Years," the Western political elite has tried to close its eyes to the depth of the Western crisis, the seriousness of the actual general low economic groups (as compared to the golden years), and firmly believes in "change." Politicians and experts persist in advocating for change, whereas transformation is needed instead. Few transformations have taken place, unfortunately. This is probably one cause of the rising anxiety in our societies (Gutmann, 1989).

An exception, during the 1980s, was the process of transformation that took place in Great Britain, under the leadership of Margaret Thatcher; however, elsewhere in Europe and even in the United States or Japan, no real political impulse has succeeded. Although this process of establishing systematic economic liberalism in sectors of the British economy has generated a large number of poor people, there is no doubt that Great Britain, founder of the welfare state in Europe, has gone through a transformation process by setting up a new structural capitalistic society. As confirmation of the existence of this transformation process under Thatcher's leadership (and John Major's), which was a progression *and* a regression for Great Britain, let us note that it is Tony Blair's *New* Labor party which was brought to power in 1997, heralding a new era.

The desire for transformation is an expression of the vitality of any institution or system, whatever it may be. It is both the result of and a condition for the viability of social innovation. On this particular point, *desire* seems to be a prohibited word in modern society. It is associated with a form of guilt. Moreover, desire in its expression is linked to an image of death (because of AIDS, sexual delinquency). However, *desire is also life.* Many institutions—whether firms, political organizations, or social institutions—do not take this into account. They are more preoccupied with surviving than with living an enriched life. By institutional transformation, we have faith that institutions will not only perpetuate, but also innovate accepting through their leaders both the face and the fate of transformation—that is, incompletion. For an individual, to live means also accepting death. In the same way, *one of the qualities of the leader is acceptance of the incompletion of transformation.*

## III. GENERATIVE AND GENEROUS LEADERSHIP: A CONDITION FOR TRANSFORMATION OF THE INSTITUTION

It has become very difficult to think about leadership today—probably because of what we learned about or experienced with some leaders earlier in the 20th century. The traditional perspective for studying leadership is in relation to the theory of the crowd elaborated on by Freud (1921/1981). According to this theory, the crowd is a unique functioning entity that interacts with one individual (the leader) in a fascination-domination relationship. What holds individuals together in a crowd is the belief in a supreme leader who loves each of them equally. In the Catholic Church or the Army, for example, the link between each individual and the supreme figure (Christ, the commander, etc.) creates mutual dependencies. Finally, what holds the institution of the crowd is a libidinal link between members and the supreme figure. The leader is in this case the ego ideal, the superego (Freud, 1923/1989). There is no doubt that in huge corporate organizations, for example, the leader tends to be at the center of a strong political, unconscious, and spiritual game.

What is a leader? A leader is a visionary. He or she stands out and is followed. In our experience as consultants (more specifically, as advisers in leadership), we would like the leader to have seven characteristics (Figure 3). Of course, these features can be found in persons other than leaders, but only leaders can incorporate all of them and adapt them to various circumstances. Our view is that *leadership has to be generative.*

Figure 3. The characteristics of leadership.

First, *vision* is the most obvious quality that a leader must have. Using an example in the political field—de Gaulle's departure to London in 1940 and his speech of June 18, which gave birth to the French Resistance, were visionary. Churchill's absolute refusal to compromise with the German Nazis after 1939 was visionary in the sense that very few would have thought that this was the right attitude at the time. The leader must perceive the primary task of the institution he or she belongs to and conduct the institution according to this objective. With vision, the leader will be able to set up an institution following a more or less unknown path, and to announce the "B" desired and fantasized task, which is the next step of transformation.

For example, progress of medical and biological technologies is an issue in which vision is urgent, since mankind has today the possibility of interfering with human creation. Any hesitation in our ethics about research or experimenting on human organisms today could generate disasters tomorrow (see chaos theory). Concerning the bioethics issue, no one is qualified to tell what rules have to be adopted: Options and opinions are too divergent, since cultural practices, beliefs, and nationalities can influence the individual's opinion. However, possible scenarios have to be clarified: We have to know what can happen to humankind in relation to these biological and medical revolutions. In a sense, vision is required to make the right choices in terms of regulation. In this instance, vision is a condition for ethics ("ethics" being understood as "the art of directing one's behavior"), since it helps individually and collectively to follow a clarified or, at least, voluntary path.

The construction of the European Union after World War II is another example of vision, both in its economic and political aspects. In the early fifties, French leaders Robert Schuman (minister of Foreign Affairs) and Jean Monnet (the "Father of Europe"), along with political figures such as Konrad Adenauer (German Chancellor) and Alcide De Gasperi (Italian Prime Minister), set the foundations of the actual European Union. The movement toward the EU was a succession of progressions and regressions. Still, there are behind this movement true visionaries. The vision was carried out by political leaders like Helmut Kohl, German Chancellor, and François Mitterrand, the French President, who set the basis for the single currency.

A second feature of the leader—*discernment*—implies that the leader must be constantly aware of clarifying his or her internal and external environment by taking into account the economic, political, psychic, and spiritual issues that are at stake within the group. Discernment is *staying with one's desire.* Because leadership involves a considerable output of energy, a leader has to convince people and fight resistance. In a sense, discernment is to the leader quite similar to what the "sublimation process" is to the artist—that is, to give way to desire in a more or less socially acceptable form.

Let us think about how today's high-tech economic successes, such as Microsoft or Apple (which was still thriving several years ago) started in the 1970s: Teenagers fascinated by computer science spent endless hours at their computers. At the core of their task was a desire for innovation and transformation. Discernment led to vision. Another example, George Lucas's insistence on licensing the derived products of his film *Star Wars* in 1977 not only made him rich, but also generated a model of merchandising that is now systematically used by the entire film industry in Hollywood. His decision was strongly criticized at the time.

Third, the leader must have *a capacity to have followers*. Followers play an active role for creative leaders and are part of the institutional transformation. In our sense, they must carry on an "authoritative followership," which means that their support for the primary task of the institution is *deliberate and chosen*. This condition distinguishes a totalitarian system or a dictatorial regime, in which people are forced into following, from other systems. An "authoritative followership" is a relationship with the leader that allows each individual to be an author, or more precisely a coauthor, and is also a relationship in which the leader seeks systematically to enable acts of authority from followers. A leader who has lost the passionate sense of mission for his or her people cannot continue to lead them, for he or she cannot enable them to become coauthors. Moses leading his people to the Promised Land is an example of the missionary love a leader must have for his followers. In contrast to "authoritative followership," an omnipotent leadership is based on a dependency relationship between followers and leader. This dependency structure is recognizable in many of the firms in which the leader is also the founder. It hinders the capacity of other managers or employees to use their authority to develop vision in the firm.

One great responsibility of the leader toward followers is his or her constant effort to clarify and abide by both the internal and external boundaries of the institution. It is especially important to enforce boundaries between public and private matters. A leader must be aware of boundary confusion, since this systematically affects the primary task of the institution—for example, a manager using the firm's resources for personal purposes, or a sect-member employee preaching to colleagues.

Above all, a leader must be able to set up a *working pair*. The leader must find a manager—a person whose mission is to prepare and organize the means and resources necessary for the primary task of the institution. Any institution needs then two roles—usually filled by two different individuals. In a way, the manager's role is to implement in daily life the vision of the leader. *Management has to be shared.* An important role of the manager is to help members of the institution to "manage themselves." Again, this is the process of authoritative followership, which implies that the leader's followers have authority over themselves and their activities within the institution. Such a process allows better efficiency in achieving the primary task of the institution.

Unfortunately, working pairs are uncommon in both the economic and the political world. An example of a working pair in the mid-1990s, which fills the business world with admiration, is the case of Bill Gates, President of Microsoft, and his Executive Director, Steven Ballmer (Executive Vice President of Sales and Support). In this pair, Gates takes the role of the visionary leader and Ballmer of the tactician manager. Seven key executives back them up: Not only is this working pair effective, but it offers an example of shared management. In the political world, the building of the EU (toward the single currency) during the last decade was essentially given its impetus by French and German leaders (Mitterrand and Kohl), but their vision was adapted by great civil servants such as Jacques Delors (at the head of the European Commission), who in this case played a managerial role.

Additionally, the leader must create and share authority within the institution, and therefore contribute to the grooming of potential leaders to take his or her place, the attribute of *generativity/generosity* (fourth characteristic). It is imperative that a leader succeed in engineering his or her *succession:* Too many firms have collapsed because of failed successions. This has occurred most frequently with firms whose corporate leaders had forceful personalities. This is also a crucial economic issue in small and medium enterprises. Often, such institutions depend on the personality and energy of their founder. When the leader's departure is at stake, the tension and anxiety (not contained anymore) about succession is so strong that the institutional infrastructure may fall apart.

Several strategies about the succession of a leader are usually at work in an institution or firm. The most obvious strategy is conducted by "presumed successors"—the individuals who form a "court" surrounding the leader. Always agreeing to the leader's decisions, they form the nucleus of the leader's supporters. They are the legitimate successors. The second strategy is used by the traditional opponents to the leader in the firm, who exist in any institution. They form a group that represents opposition to the leader's decisions and methods of ruling the firm. They are tolerated and often secretly admired, since this faction counterbalances the leader's power and also serves as a medium to express the dissatisfaction of the members of the institution. This group can serve as a good barometer of opinion for the leader, also. Third, a group of outsiders have their own strategy in the succession process. The "outsiders" play a role perceived in the institution as very individualistic: the clown, the genius inventor, the clerk, and so on. They are perceived by the others as not having integrated the institutional logic or collective game and are therefore cast aside. Each member of these subsystems is a potential leader. The model described here is a caricature of real strategies for succession. Careful observation of the collective representations and roles in the institution can always reveal such behaviors.

We are very concerned about the question of *how the leader can generate other leaders,* for we observed that many failed successions implied the failure of the institution, probably caused by the tremendous anxiety that emerges during the crisis created by the departure of the leader. In our experience, in order to achieve the succession, the leader's successor must be carefully selected. The potential leader must not possess characteristics deleterious to the future institution.

The difficulties of the succession process are similar to the difficulties of innovation. In a sense, to innovate implies betraying psychically one's parents (Mendel, 1992). A successful succession is also a form of *betrayal.* To explain this by a metaphor, one has to escape from the prison created by the shadow of one's predecessor. Similarly, each individual is confronted with succeeding his or her own parents. Within an institution, the potential successor is confronted by the image of his or her parent in the person of the predecessor. The successor has to "betray" the predecessor psychically, in order to achieve the succession. An example of this phenomenon is the case of a brilliant young civil servant nominated as head of one of the most prestigious universities in France. He took over from a well-known elderly director.

The new director was nominated to a very important function (the equivalent of the U.S. Supreme Court). One of his first decisions was to allow first-year students a second chance on their final exams—a measure that had been refused for years by the former director. Long considered a student of the former director in the university, the newly nominated leader thus symbolically killed his father. For this one successful succession, how many other organizational failures plunge institutions into turbulence?

The greatest difficulty in successions is to avoid repetition and achieve innovation that aids the institution. To show how institutions can be involved in repetition processes just like individuals, let us take the case of the Italian car industry, dominated by the Agnelli Family. The empire was founded at the end of the 19th century by Giovanni Agnelli, *"il Senatore,"* who never thought that his son, Edoardo, would succeed him. The son eventually died in a plane accident. Instead, he appointed his grandson, Giovanni Agnelli (the *Avvocato*) to head the Fiat firm. Now that the *Avvocato's* own succession is at stake, all signs indicate that he has chosen his nephew Giovanni, the executive chairman of Piaggio, to succeed him, which is probably a repetition process. In fact, in Italian, the same word is used for the grandson and the nephew *(nipote)!*

A leader, when the time of succession approaches, has a tendency to concentrate even more power in his or her own hands, often encroaching on the role of manager or even consultant. But by putting himself or herself in another role, the leader cannot be generative. Unfortunately, leaders often raise—consciously or unconsciously—confusion between the roles, either by increasing their monitoring of the administration of the firm (being a manager), or by leaving its primary task and acting as a fake consultant.

In a context of crisis—whether it is caused by external factors (tough competition, oil crisis, political instability, etc.) or by internal factors (from the coming departure of the boss to an employees' strike)—leaders typically start confusing roles. Probably, crisis is an environment in which role boundaries are difficult to distinguish, to establish, or to accept. But this is obviously a time when boundaries should be carefully worked through and then, once achieved, strongly supported and sustained.

The fifth characteristic of the leader—*appetite*—is the tendency of the individual to satisfy his or her needs and desires. This occurs in every individual, but in the leader is particularly acute. Unlike a depressed individual who has difficulty acknowledging his or her own desire, a leader must have appetite for himself or herself as well as for leading others. On the other hand, the leader has to realize that his or her appetite will never be totally satisfied, which is related to containment, the next characteristic.

The process of taking charge of the destructive part of the group's anxiety can be called *containment*. The leader must have a strong capacity to contain the dissatisfaction of the members of the institution, as well as his or her own dissatisfaction. As transformation of the institution is always in process, the leader must contain the state of incompletion. Noncontainment of anxiety and violence implies that dissatisfaction can emerge any time. A good example of this is the 1992 Los Angeles riots or the civil rights protests in the United States in the 1960s. In both of these cases there was an absence of recognized leaders within both the establishment and the community; riots broke out because no one was "in charge"—that is, in charge of containment. Noncontainment of negative feelings and affects can also create overwhelming disorder within the institution.

Therefore, the institution and its members need to rely upon the leader and his or her ability to contain. The relationship between them is one of *dependability* rather than dependency. Additionally, authoritative followership and shared management allow the leader to rely on followers. It is a two-way dependability.

Finally, the leader is a person of *path*. In a changing, unstable environment, in which any attempt to achieve a defined project inevitably fails, the leader has to maintain a steady path. This *path* is a translation of the leader's vision, whereas the *project* is a reduction of this vision. The leader on path has the capacity to live, to carry, and to bring *surprise* to the others. For only surprise can enable the leader to escape from the hold of the institution. This "hold" is implemented by the system and its environment in order to put an end to the ongoing transformation and generate a process of repetition.

The leader of an institution is writing the members' biographies, by giving a meaning to their own path. By doing this, the leader also writes his or her own biography, and forges his or her own path. This may be the reason why it is important to study leadership "inside" the transformation process. It is less important to emphasize the *qualities* of the leader than the *relations* of the leader with the other members of the institution, in the context of transformation.

## IV. CONSULTATION AND INSTITUTIONAL TRANSFORMATION: THE SUSTAINING TRIANGLE

The *working pair* that has to be introduced in the practice of leadership is fundamental: It enables each role in the institution to be clearly defined. The *leader* defines his or her vision, and is generous and generative (accepting that other leaders diverge from his or her path). And the *manager* tries to create and organize the means of fulfilling the primary task of the institution as well as enabling members of the institution to manage themselves. However, another role must also be introduced into the dynamics of the institution: that of the *consultant*.

We call the threesome composed by the leader, the manager, and the consultant the *sustaining triangle* (see Figure 4). This is a conceptual archetype which represents the conscious and unconscious interactions that can take place between the three roles carried by either individuals or by groups, in order to maintain an equilibrium within the institution and to foster institutional transformation.

We cited Microsoft, with Bill Gates and Steve Ballmer as an example of the working pair. The head of the computer giant also offers a good example of the sustaining triangle, if we consider the role fulfilled by Nathan Myhrvold, Technology Director of the company. His close relationship with Gates as well as his multidisciplinary profile (computer expert, doctorates in physics

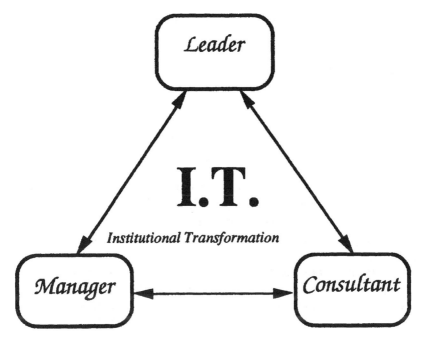

Figure 4. The sustaining triangle.

and mathematics), his numerous passions (paleontology, cosmology, fishing, racing, photography), and his reputation as inspired visionary make him an effective consultant. Many of his notes have influenced Gates's strategy over the last few years. The sustaining triangle at the top of Microsoft would then be interpreted as Gates as leader, Ballmer as manager, and Myhrvold as internal consultant.

The consultant's role is necessary both for the members of the institution and for its environment, because the consultant represents the Other, in bringing diversity and otherness. He or she thus plays a role similar to the "mirror" of the ego of each individual as defined by Lacan (1965). According to Tournier (1972), Robinson Crusoe states after his encounter with Vendredi: "The Other is a powerful element of distraction for us, not only because he bothers us continuously and keeps our mind busy, but also because just the possibility of his appearance throws a vague glimmer on objects of our universe which were before located in the margins of our attention, but can become its center at any moment" (p. 262). The consultant's role cannot be better defined.

In reality, the consultant breaks the potential one-to-one totalitarian relationship between leader and manager, completing and enriching it. Ideally, he or she can therefore be the spokesperson for what is rarely or never expressed within the institution, and for the members who rarely or never speak. The consultant's intervention is naturally conflictual. Thus, the consultant's role or figure enables a permanent fight against leaders' temptations to consider themselves as all-knowing and all-powerful.

An important state-owned French company was the theater for such a scenario at the end of 1996, when the chairman decided to dismiss the four managing directors and replace them with an executive counsel of seven members over whom he would preside. The former CEO and vice-CEO became vice-presidents, losing their "operational" posts. Changing or reorganizing the management of a company is usual, and choosing managers is also an understandable task for a chairman. But this particular decision was particularly amateurish and autocratic for several reasons. First, the company was legally set up at its founding so that it would be ruled by both a chairman and a CEO, a dualist form of government. Therefore, the chairman's decision was invalid. Most importantly, the conditions under which the decision was made and announced to the CEOs, outside the jurisdiction of any internal institution such as the board, was completely inappropriate. The entire process was carried out in a manner in total disharmony with the practices of a modern company, but right in tune with the autocratic temper of an incompetent chairman. As a result, the chairman lost credibility with his employees. The French government then had to intervene, maintaining the post of CEO and giving it to the former vice-CEO.

At the head of this important French company, the sustaining triangle is now composed of a completely different structure. The actual CEO is assisted by two managing directors. It offers the possibility of a threesome in which the CEO is the leader and the two managing directors the managing entity, with an adviser in leadership to complete the triangle. Additionally, each member can evolve from one role to another (leader, manager, consultant).

This sustaining triangle is a conceptual archetype: It is necessary that each person's role in the institution evolve. First, each person has the capacity to be at once a leader, a manager, and a consultant. These different roles are carried out by each person according to circumstances, in interaction with the others. Second, everyone is able, at least theoretically, to fill each of the three roles. Additionally, each role can be more or less present at the same time in either an individual or in a group (Figure 5). Sometimes, one of these roles tends to prevail over or be exclusive of the other roles.

The "Institutional Transformation" working conferences organized by members of Praxis International and by the International Forum for Social Innovation (IFSI) are an example of this exchange of roles within the sustaining triangle. In these conferences, the staff fulfills two roles. First, it acts collectively as management of the conference. It thus assumes the responsibility for managing the boundaries in such a way that members may confront the primary task of the conference. Second, staff members also intervene as consultants during events. The staff in its interaction with the members must be absorbed by either role. For example, if staff members focus on the role of management, they might not fully offer their own perception and experience of events as they occur during the sessions, which eventually hinders social innovation. Similarly, in these working conferences, the director no longer takes the role of consultant in the Large Study System—that is, he does not attend the Large Study System session. This allows him to work on his role of leader. Not being a consultant, he can more effectively carry out the leader's role, for himself and for others, inside the temporary institution of the conference.

The principle around which we tend to organize our institutional transformation working conferences is that *being eternally enlisted in one role can be a source of suffering, entropy, and impotence.* During different periods of our lives, we must carry out different roles. Many consultants feel the need to become, after several years of practice, a manager, or a leader in an entrepreneurial role, and create their own firm.

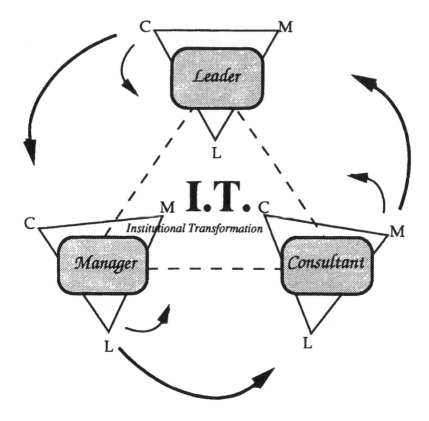

Figure 5. Evolving roles in the sustaining triangle (L = leader; M = manager; C = consultant).

Finally, we want to insist on the role of the consultant—adviser in leadership—in the institution. Advisers in leadership build a long-term relationship with executives in order to apprehend the evolution of their role. The aim of this partnership is institutional transformation. The efficiency elicited as a result of such a process eventually benefits the entire institution.

Advisers in leadership are not experts, but enlighteners: They bring a different and objective perspective to a decision, role, system, or strategy in an institution. With their partner, they can thus share amazing renewed vision of the institution. This enables leaders and managers to emerge from collective inference processes and lead action in a different manner.

In a sense, advisers in leadership are the complement to the leader's or manager's loneliness and isolation. There are some decisions that an individual at the head of an institution is forced to make alone. The consultant is there to compensate for this particular characteristic of leadership.

To conclude, any living and not just "surviving" institution should create an internal environment where management is shared with as many people as possible, and where leadership is not only creative, but also generative. It should be generative of ideas, goods and services, of followers with authority, and, finally, of new leaders. A successful succession is necessary for the existence and vitality of the institution. The presence of a consultant does not create a miracle, but he or she can help to prevent the managerial search for omnipotence and eternity in the successor. Aiding in this transformation, the consultant will accompany the differentiation of roles by being part of the sustainable triangle of leader, manager, and consultant.

We would like to finish with another example of a failed leadership due to an impossible transformation: the "Club Mediterranée." A few years ago, Serge Trigano succeeded his father Gilbert Trigano, founder of the firm. The group was confronted by several difficulties, and in spring 1997, the stockholders placed a new chairman in the company: the successful chairman of EuroDisney—Philippe Bourguignon. Our working hypothesis is that the son (Serge) refused to acknowledge his differences with his father (Gilbert). Gilbert was both a leader (his vision of "a new kind of holiday" led him to create the Club Mediterranée) and a manager. First, Serge refused to see that he could only succeed him in the role of manager during his training periods. Second, his leadership should have been realized by other means than his father's influence, such as developing his own personal skills. Serge may have been conflicted by unconscious drives that prevented him from adhering to his desire, from psychically "betraying" his father. His refusal of a consultant (an adviser in leadership) also reflected this state of mind. It may be that in this process of succession, family and entrepreneurial dimensions were confused by Serge Trigano, and that this confusion altered his conquest of autonomy as a leader. In a sense, the transformation process was refused, and instead a conformation process was achieved . . . until Serge Trigano's dismissal by stockholders.

Coming back to the example with which we began this chapter, conformation (or "cloning") cannot be the solution for leadership, whether an individual heads an organization or is an employee. Leadership is effective when one is utilizing full authority and displaying to some degree the seven attributes previously discussed. It cannot be separated from a transformation process in which each individual is coauthor. Innovation should allow transformation. Each of us has the capacity of being coauthor of institutional transformation. This entails the capacity to envision options other than repetition and reproduction of past behaviors; it also implies the ability to transform roles, relations, systems and, most of all—projections!

## REFERENCES

Dahan Dalmedico, A., Chabert, J.-L, & Chemia, K. (1992). *Chaos et déterminisme,* collect "Points Inedit Sciences." Paris: Editions du Seuil.
Damasio, A. (1995). *L'erreur de Descartes ou la raison des émotions.* Odile Jacob.
Foucault, M. (1966). *Les mots et les choses.* Paris: Gallimard.
Foucault, M. (1969). *The archeology of knowledge.* Paris: Gallimard.
Freud, S. (1981). Psychologie des foules et analyse du moi. *Essais de psychanalyse* (pp. 41–115). Payot. (Original work published 1921)
Freud, S. (1989). *Au delà du principe de plaisir.* Paris: PUF. (Original work published 1920)
Freud, S. (1989). Deuil et mélancolie. *Oeuvres complètes* (pp. 259–278). Paris: PUF. (Original work published 1917)
Freud, S. (1989). "Le Moi et le Ca." *Uvres complètes* (Vol. XVI, pp. 255–301). Paris: PUF. (Original work published 1923)
Girard, R. (1961). *Mensonge romantique et vérité romanesque.* Paris: Grasset.
Girard, R. (1972). *La violence et le sacré.* Paris: Grasset.
Girard, R. (1978). *Des choses cachées depuis la fondation du monde.* Paris: Grasset.
Gutmann, D. (1988). The decline of traditional defenses against anxiety. *Proceedings of the First International Symposium on Group Relations* (pp. 5–22). Keble College, Oxford: A.K. Rice Institute, 1988.

Jullien, F. (1992). *La propension des choses; pour une histoire de l'eficacité en Chine*. Paris: Seuil.
Lacan, J. (1965). Le stade du miroir comme formateur de la fonction du Je telle qu'elle nous est révélée dans l'experience psychanalytique. *Ecrits* (pp. 93–100). Paris: Seuil.
Mendel, G. (1992). *La société n'est pas une famille*. Paris: Editions La Découverte.
Schumpeter, J. A. (1939). *Business cycles: A theoretical, historical and statistical analysis of the capitalist process*. New York: McGraw-Hill.
Tournier, M. (1972). *Vendredi ou les limbes du Pacifique*. Paris: Gallimard.

# 2

# The Changing Psychological Contract in the Workplace

*Boris M. Astrachan, M.D., and Joseph H. Astrachan, Ph.D.*

In a world in which authority is legitimate and just, loyal obedience is a duty (Schaar, 1970). This is the idealized relationship of man to God, of liege-man to his king, of soldier to officer, and even of contented employee to benevolent manager. Yet even in the idealized relationship of authority to a follower—of God and man or employer and employee, there is the assumption of a mutuality of interest. In exchange for adoration, service, or the performance of work, those pledging service will be cared for, protected as well as possible, and rewarded, if not in this world, then in the next.

Yet the world has never been perfect. The relationship between leaders and followers, between organizations and members is continually in flux. These relationships are subject to changing environmental forces, to the need for and availability of workers, to agreement about the extent to which values are shared, and to the existence of some mutuality of interest.

How generalizable is the silent contract of mutual obligation in the workplace? How has the relationship between leaders and worker compared to the ideal? More often than not, the work-management relationship has been conflictual or oppressive.

To examine the conditions under which a strong psychological contract might flourish, we will first look at the health care industry. Next, we will look at other settings, attempting to conclude with some overview of the conditions that are necessary to support a strong psychological contract between organizations and their employees. Then we will identify some of those issues that act to destroy trust and the psychological contract. Finally, we will identify some issues in the world of work that must be addressed if we are to reestablish (or establish) the psychological contract.

## HEALTH CARE

Until quite recently, the health care industry has had three major groups involved in organizations: boards and their delegates, managers, and physicians—and everybody else. For over 100 years, these groups vied for control, not only over institutional direction, but also over the lives of those who labored in these institutions. Rosenberg (1987) describes the extraordinary control of the life of nurses, orderlies, and junior physicians resident in the middle of the 19th century. He quotes from the diary of John Duffe, a ward nurse in 1844. Indignant because of the rotten butter served to nurses and patients at the hospital, Duffe wrote, "It is useless to complain because if we do it may irritate the religious feelings of our superintendent and he will soon point out the way to the gate." Duffe goes on to note that one must eat rotten food, "for like everything else if we complain our godly superintendent will tell us you may consider yourself discharged for the Lord sent it and you must eat it you hireling" (pp. 43–44).

Staff could not even leave the grounds without permission, and a day off was not granted unless someone else (often a patient) was found to do the work. Hospitals and health care settings were agents of charity, and those who required services there (particularly those who could not pay), the staff, and often the junior physicians were subject to intensive supervision and control. In fact, even as late as 1923, an intern could not marry another house officer. Activities to limit the workweek of junior physicians to a maximum of 80 hours per week would have been viewed as an unattainable luxury.

Throughout the 19th century and into the 20th, the image of the hospital was one "of an enlarged concept of stewardship organized around a vision of the common good—one in which a potential conflict between interest and Christianity could not exist. The stewards of society's worldly goods were necessarily the proper stewards for the interests of their community's less fortunate members" (Rosenberg, 1987, p. 105). Authority spoke for the good of all, and one either obeyed or was branded immoral, was discharged, or worse.

One of the early reformers of the hospital system was Florence Nightingale. Her role in beginning the professionalization of nursing is well known. Perhaps more important was her influence on helping to create an orderly, disciplined hospital with nursing engaged in this moral and healing task (Rosenberg, 1987, pp. 128–138). Nursing training emphasized hard work, discipline, and moral order, as well as more technical tasks.

Nursing students near the turn of the century "worked a minimum of sixty, and more frequently, seventy hours a week on the wards [and] . . . accepted a discipline that could not easily have been imposed on an equal number of men" (Rosenberg, 1987, p. 221)—largely in the context of what we today perceive as the sexist exploitation by paternalistic males of deferential females, accentuated by the threat of expulsion and loss of livelihood for disobedience. However, nursing was one of the few occupations available to women in which there was the hope of an acceptable work life. Thus, nurses, then as well as now, endured much and struggled for some share of security and even authority.

As hospitals became increasingly professionalized and became a setting for professional training, the experience of nonprofessional staff was even more different from that of doctors, nurses, and managers. "They adjusted to authority rather than exerted it. Work conditions often deteriorated as the paternalism that had traditionally tempered bleak realities was gradually transformed into a self-conscious and impersonally bureaucratic style of management" (Rosenberg, 1987, p. 282).

Since the end of World War II, there has been dramatic change in work relationships within hospitals, including the growth of professional management, which altered interprofessional work relationships, and generated substantial challenge to the autonomy of physicians. While professional groups have been engaged in a struggle in order to achieve some redefinition of authority relationships, other workers—and even nurses—have moved to unionization and direct confrontations with management in order to attain security, improved working conditions, and enhanced status.

Even a cursory reading of the history of health care professionals suggests that the psychological contract between employer and employee is not now—and has never been—stable. In this world, some groups, physicians primarily, have been privileged and irreplaceable, and others have been viewed largely as interchangeable cogs (e.g., attendants, cleaning and maintenance staff, food service). The senior physicians, managers, and, to a lesser extent, nurses, following extensive periods of apprenticeship, had some sense of place, of their importance in the scheme of things, and felt for much of the time sufficiently well taken care of to return loyalty.

In health care, and in industries whose histories extend back over 100 years, the silent psychological contract was originally a limited one. More elite groups—those whose values, beliefs, class, religious identification, and racial and gender characteristics were identical to those of organizational sponsors—might develop their skills in the organization and in exchange for loyal, hard work might prosper and attain a measure of security. Over time, other groups sought to enter this privileged status—to be able to, in exchange for their loyal, hard work, aspire to security and even some measure of authority. The majority of workers, the lower level employees, were largely discounted, and subjected to oppressive treatment and to immediate dismissal for challenging authority.

Over the past 10 years the organization and corporatization of health care has disrupted the silent contract, even for elites. The restructuring of health care has dramatically changed practice. To enhance their competitiveness, hospitals have engaged in operations improvement activities. These have led to changed procedures, the substitution of technicians for professional staff, and staff reductions. There have been hospital mergers, linkages of hospitals into networks, and the development of hospital-medical group organizations. Health care has become corporatized (Starr, 1982). In this process, the functioning of elite institutions such as academic medical centers had been threatened (Iglehart, 1995; Kassvier, 1995). The market has changed the way physicians are paid, and a renewed emphasis on primary care has replaced the priority given subspecialty care. For the first time, some physicians are having difficulty finding positions, others are seeing their income reduced, and still others are losing their positions altogether (Leigh, 1994; Rivo, Mays, Katzoff, & Kindig, 1995). Some health care policy advocates call for radical approaches to limiting physician numbers include limiting access of foreign graduates to residency training (Whitcomb, 1995). Not only is the silent contract in industrial settings threatened, so too are implicit societal contracts regarding health care.

## OUTSIDE OF HEALTH CARE

There are corporations today which insist that they and their employees are involved in a shared enterprise—in a moral endeavor. They speak to a commitment to excellent work, to high standards in production, to innovation and creativity, to excellent service, and to the maintenance of excellent relationships with their clients and their employees. In many of these corporations, the employees share the values of management.

The benefits guide of one such corporation extoled the reciprocal relationship of company and employees. It insisted that as the employees serve the company, their loyalty is rewarded by the equal dedication of the company. The benefits to employees enumerated range from a commitment to full employment, a merit promotional and pay system, and open communications from lowest to highest levels, as well as a generous benefits package.

The commitment illustrated here was not rhetorical. It was a real and a lived part of corporate life. This type of commitment often reflects a dedication to a particular style of management, often identified as "theory Y" (McGregor, 1960), which emphasizes open discussion and joint decision making, and which assumes that if you expect employees to be responsible and creative, they will be.

In the late 1960s and early 1970s, one of the chapter's authors (BMA) was part of an interdepartmental seminar on organization at Yale University, along with Clayton Alderfer, Portia Bowers, Al Fitz, Jim Miller, and Ed Klein. In order to further the work of exploring together various theoretical constructs about organizations, the group offered its skills as a consulting team to a large corporation (Alderfer & Klein, 1978). The group asked for the opportunity to study a "model plant, and in exchange for their paying our expenses and allowing us entry, we offered to report on issues and potential problems that we identified." In 1970 the group was invited to consult in such a plant. Within the corporation, this plant was viewed as an important organizational experiment. The managers were selected for this work not only because of their technical competency, but also because they all believed, and acted on the belief, that individuals who are treated as responsible and self-motivating will behave responsibly, be self-directed and creative, and feel a sense of ownership of the organization.

In the investigation it was clear that the plant lived out its ideology, and that it was highly successful in the corporation. It was one of the two plants producing products of use to large sectors of the corporation. It could deliver its product more cost effectively and with much less spoilage than its "sister" competitor. It could do so even to plants located hundreds of miles away, in the immediate proximity to its "sister" competitor.

Most of us know that "theory Y" management, that is, shared, open, democratic management, works well in relatively small (50–200 people) technical industries, where there is a motivated, professional, primarily white-collar work force with extensive shared values about responsibility and work, and a sense of community (Alderfer & Klein, 1978). However, this was a manufacturing plant, composed largely of blue-collar employees, in a unionized industry. This plant was an anomaly, resistant to union organization. Employees, in private sessions with the consultant group, were vigorous in their opposition to unions. They espoused traditional small-town values, insisting that good work ought to be the basis of reward, and not only seniority. They believed that innovation and accomplishment merited reward, and that pride in work was important. They valued the extensive cross-training that was routine in this plant. They believed in pay comparable to that of workers in similar jobs in their geographic area. They knew they were paid less than employees in the unionized "sister" plant, and acknowledged that unionization might result in increased income. But, by and large, they saw unionization as a constraint on both management and individual employees. One employee went so far as to insist that unions were the response of workers to autocratic, inattentive management.

The corporation was, of course, very interested in this plant because of the high morale, high productivity, and reduced costs of operation. It arranged to have bright young managers rotated through on a regular basis. The consultant group was particularly interested in the implicit psychological contract between workers and managers, and sought to study this relationship. A "humanistic paternalism" was identified as central to this management-employee relationship. Managers articulated overarching values with which most employees agreed. Traditional values, emphasizing responsibility, pride in work, and pride in self, were emphasized. There was, as well, emphasis on communal values—a commitment to each other, to family, to community, and to stability. Managers personified these values and were clearly committed to their employees, to the plant, to its products, and to the larger organization. This is characteristic of small town (or small city) life, where resources are scarce and individuals interact with each other in multiple domains (e.g., work, church, clubs, markets) (Astrachan, 1988). A major challenge for industries is how to provide some of the conditions that permit such value sharing in larger urban settings.

In addition, managers were also technically skilled and dedicated to the work. Most importantly, managers were able to protect the plant within the context of the larger corporation. They could protect innovation and difference. They insisted that if new labor-saving technologies were developed at the plant, no one would lose a job. They guaranteed full employment and challenging work. It is no surprise that they were believed and trusted.

Employees largely shared management values. New hires were all employed in the least technical part of the organization. There they learned work skills and were continually oriented to plant values by managers and more senior employees. Employees who did not "fit in" never seemed to get out of that section of the plant.

The secret psychological contract was alive because managers and employees really were supportive of each other. They shared values, and they were protected from adverse environmental forces, and even to some extent from adverse corporate actions.

## CHARACTERISTICS OF A SHARED SILENT CONTRACT

In order for a shared psychological contract to exist, workers and managers must agree on the nature and extent of their interdependency. There must be some shared values. While we have so far emphasized values stressing interdependency and commitment, pride in work, and striving for personal responsibility and personal esteem at work and elsewhere, these values have not been historically shared, and indeed entry into interdependent relationships has been denied to many individuals on the basis of race, religion, ethnicity, or gender.

The world in which we live does not remain constant. The 1960s and 1970s were a period when we as a nation began to more actively confront racism and sexism, and when it became clear that the old "moral" order countenanced discrimination, injustice, and immorality. In the 1980s and 1990s we have not yet effectively integrated the mass of the newly enfranchised into the work force. Organizations and individuals may ask themselves, are their values our values? Can women be trusted, or Blacks, or Hispanics? The failure to adapt to new people, and even new values, perplexes our organizations, and helps foster alienation from the organization. Whose organization is it anyway? What does it mean to me to have to compete with them? Many workers who thought they represented the organization have had to adapt to new working conditions, and for many the process was difficult. Many employees became increasingly focused on the self. We saw the growth of the Yuppie generation, of making it big, of being a "hitter" or "player," with little concern for others, "the little people." As athletes compete for multimillion dollar contracts, no one seems to care about what salary ticket clerks, trainers, ushers, and locker room attendants earn. It is an era where young people make millions manipulating corporate takeovers without ever seeming to comprehend that peoples' lives and livelihoods were manipulated, even destroyed. It is an era when trusted managers might become powerless to protect their employees. In all industry, including health care, restructuring with a focus on reduced costs, increased accountability, and decreased security for employees has become common (Samuelson, 1995). The world of work seems threatened by mergers, cutbacks, layoffs, and the pain associated with these changes in the work world.

Astrachan (1991) was particularly interested in individual responses to experiences of mergers and acquisitions. Listen to the destruction of a sense of security and of commitment in this passage: "While observing an individual whose company had just acquired a sizeable competitor, I was perplexed by the degree to which he and his associates were actively denying the potential hazards of their situation. . . . I asked if they knew that their firm planned to lay off at least 6,000 employees as a result of the acquisition. I was then told the rumor that 10,000 would be let go" (p. 4).

Harshbarger (1987) relates the following illustration: "On the Tuesday before Thanksgiving we were shocked to learn that Sealy, Inc. had just been sold to the Ohio Mattress Company. Happy Thanksgiving. By the end of December, the company had undergone a takeover. Merry Christmas. My senior colleagues and I greeted the New Year with foreboding and with good reason: by Mid-March, most of us were unemployed" (p. 4).

In another company, pervasive weight gain in employees was noted following announcement of a takeover, and there was also significant increase in cigarette sales (Berenbeim, 1985). Fear, graveyard humor, shock, and disbelief are all common. A senior executive caught up in a major restructuring reported, "Many of us feel we have been loyal to something which no longer exists. Perhaps we were in love with our own loyalty." The sense of connection is ripped. The contract violated.

For many, the psychological contract never existed. In an exercise conducted at Yale for undergraduates, latecomers to a class were given collars and required to work under resource-poor conditions (Astrachan & Herbert, 1988). When the class ended and students were debriefed, several Black students refused to remove their collars, saying the experience was constant for them. They could not and would not remove their skin color. They felt that other students, who could walk away from the experience, needed to know that their Black classmates could not.

## LESSONS IN BUILDING TRUST

Over the last several years, we have witnessed, and even participated in, several organizational upheavals. These experiences have helped us to identify some of the major variables that build trust and that in their absence destroy the psychological contract.

### Job Security

In industry after industry, employees are treated as things, hired when absolutely needed, dismissed when bottom-line reports seem to be worsening. Labor's response is often to insist upon almost unreasonable protection, such as overstaffing and featherbedding. In the retail industry some department store chains have had severe financial problems. One high-ranking manager we know stayed with a store for the job experience, but a big part of it involved abuse stemming from corporate decisions, as well as the loss of colleagues, many of whom were financially hard hit because of job loss. To save money, lights in the store where she worked were not turned on until a half-hour before the doors opened. She often needed to arrive early, and even brought in candles in order to do company work. Her corporate loyalty, high when she started and went through an excellent training program, became nonexistent, and like so many who live through such experiences, one of her desires included "getting even." Eventually, her learning ceased and she left the company.

## Boundary Protection

To the extent that organizations are controlled by nameless, faceless others, employees are without authority, even without authority over their own work. For example, a state mental health facility with a history of strong internal direction was increasingly subject to control by regional and central office staff following a change in leadership and an announced change in mission. New programs, nominally under organizational leadership, were in fact managed at a more senior level, distant from the organization. Staff in traditional units felt devalued and diminished, senior clinicians began leaving, and care was more bureaucratized and much more difficult to access.

## History

Shared history supports shared values. The excision of the past contributes to the alienation of individuals from management. The current trend is for major corporations to lay off or force the early retirement of older workers with higher salaries, thus lowering some overhead costs. While this may also facilitate the acceptance of new directions and of new tasks, it does so at a cost. The loss of a shared history also leads to some loss in communications, because a shared vocabulary may no longer be appropriate. The organization is depersonalized by the loss of history, which further reduces motivation and commitment. In fact, what may happen is that in-groups and out-groups evolve, further splitting the organization and competing for loyalty. When bringing about major and fundamental change, it is motivationally better to acknowledge history by making it clear to all how the direction of needed changes is consistent with past actions and values. This is especially true when culture must change.

## Continuity of Communications

When programs change and management and employees are distanced from each other, the ability to communicate may be compromised. Data necessary for decision making may be lost or so delayed in transmission that work is impeded. Management and employees may perceive that their interests are separate and not shared. In an information-driven society, E-mail and faxes may inundate and overwhelm managers, and impersonal messages may substitute for dialogue.

## Reduced Intergroup Tensions

The failure to effectively incorporate minority group members and women into organizational roles often intensifies other organizational intergroup problems. Issues of gender often compound medical-nursing interaction; issues of race often complicate management-employee relationships. The failure to listen and to understand often intensifies intergroup tensions and strengthens subgroup boundaries.

## A Focus on Long-Term Goals and Reducing Recruitment of "Heavy Hitters"

There is the assumption that short-term goals require intense competition for a few "stars" who have the technical skills to build new programs or turn programs around. Extensive resources may be devoted to these few, and others may get relatively limited resources. Older employees may feel that they have been betrayed by the organization and their fields of work devalued, or they, too, may begin the process of testing out their value, their ability to be hired elsewhere as a "heavy hitter," further diminishing institutional loyalty. The process becomes even more destructive when the "heavy hitters" are perceived as having limited loyalty and yet command major resources.

A financially overextended major corporation illustrates this situation. The creditors insisted upon restructuring, and new management was installed. The new management group had no dedication to the organization or its employees. Contracts for new managers were highly favorable, with extensive protection built in should the company go into receivership. Other managerial staff and employees were convinced that the new management would be the only ones to survive, and key staff left, others became disaffected, and the expected occurred. The heavy hitters' golden parachutes saved them, but other employees, first disillusioned and later despairing, were left without protection.

## Appropriate Executive Compensation

Like the growing focus on "heavy hitters," executive compensation has become a major issue in the psychological contract. Executive salaries have exploded over the last several years and are only loosely related to company performance, as has been shown in numerous business magazines' (e.g., *Forbes, Fortune*) annual surveys of executive compensation. It is commonplace for executives to receive large annual bonuses even while thousands of employees are being laid off. The message sent to the workforce is that line work is clearly less important than administration, loyalty is not rewarded, and financial issues outweigh human ones. There is a growing emphasis on the need for employees to justify their employment and to demonstrate that they are adding value to the corporation and a diminishment of the paternalistic view that it is administration's role to find ways for workers to make maximal contributions. It is a signal that market mentality, which bases transactions on price and return, has entered into the fabric of organizational life. The conundrum is that value often requires cooperation, and a market culture reduces cooperation to short-term quid pro quo that severely limits the ability to create organizational values. Bottom-line organization leaders may not be effective in balancing short-term economic advantage and the development of organizational values promoting longer-term survival and growth.

## Addressing the Causes of Organizational Scapegoating

A challenge for any organization is the effective nurturance of staff: paying and promoting those whose work is of high quality, and educating and, if necessary, disciplining those whose work is not satisfactory. At times, organizations encounter difficulties beyond their capacity to manage. Rather than attempt complex assessment and, if necessary, do the difficult work of restructuring, the organization may identify scapegoats who carry the blame for the problem.

By its very nature, scapegoating avoids necessary work. Little is learned. At times, scapegoats may be selected because they represent values that the organization wishes to disparage. Invariably, others holding similar values feel threatened and equally devalued. Perhaps more importantly, the identification of scapegoats is usually perceived as reflecting further inequity, as disturbing the sense of psychological security.

## TRENDS THAT NEED ATTENTION

There are several trends that require attention in specific settings and in general.

First, in spite of a substantially enlarged workforce, one with many more minority group members and many more women, the values of these diverse groups have been only poorly integrated into many industries. Issues of racism and sexism continue to be common, and commonly ignored.

Alderfer, Alderfer, Tucker, and Tucker (1980) were engaged in a lengthy consultation to a major utility. Black and White managers were seen as having substantially different views of problems confronting the corporation. Minority employees saw themselves as *not* advancing, as having been restrained by racism. White male employees saw their career opportunities as impeded by unfair advantage granted to Blacks and to women.

The consultant group used an intervention focused on stimulating communication and developing contexts for representative groups to have impact on the development of corporate policy. The intervention did not fix relationships, but it helped establish a process in which issues might be honestly explored and the work environment become more supportive to all.

Second, in a world where some corporations get larger and others disappear, better ways need to be found to maintain some degree of local autonomy, so that work can be accomplished. A colleague was involved in corporate consultation in an organization where, on a monthly basis, a manager from the central office visited a particular plant. His visits were perceived as highly intrusive. However, because his technical skills were superior, and because he represented corporate headquarters, it was almost impossible to deal with his authority appropriately. His visits disrupted work, often changed directions, and eventually resulted in depression and dysfunction in one manager. The intervention was an organizational one, focused upon attempting to develop mechanisms for empowering the local managers. This intervention failed. Consultants are not always capable of making things better.

Third, to the extent that sports serve as a metaphor for our society, we perceive an emphasis on short-term objectives, on recruiting heavy hitters, on stars, and on self-serving negotiations—by owners and players. Commitment to the team and the organization is ebbing. Commitment by owners to communities is ephemeral. All of this occurs in spite of the overwhelming evidence that in order to build something lasting, one must build teams. Related to this is the trend in industry and elsewhere to pay exorbitant salaries, at best loosely related to job and company performance, to senior executives. Such behavior further saps the motivation and commitment of lower level employees, who are left feeling used and abused. Pay and other benefits of office need to be meted out in a manner that is consistent across the entire organization to build a sense of team. When the feeling that there are two or more classes of employees becomes pervasive, and two or more sets of rules becomes the norm, the fabric of the organization becomes exceedingly fragile.

Fourth, many industrial settings offer less security than in the past. Again, this has a major impact on the psychological contract, as security is the base on which relationships are constructed. As companies are merged or acquired, jobs are lost; as competition heats up, benefits are changed. When these changes occur without discussion in a trusting environment, conflict is stimulated—strikes and even bizarre behaviors take place.

One author (JHA) was part of a consulting group to a major symphony that had just gone through a serious strike. Members of the still warring factions, management and board versus musicians, initially refused to sit with one another. Routine communications were misperceived, leading to increased tension and threats. The major emphasis of the intervention was to facilitate interaction and help the diverse groups listen to each other. They were able to begin that process and to work together.

Likewise, consultations with groups aimed at helping organizations and organizational members deal with layoffs cannot eliminate the disruption and pain, but individuals can feel better about themselves, and the issues behind the process can be understood. This enables people to maintain productivity while dealing with traumatic emotional consequences.

The world of work mirrors and helps shape our society. As work is increasingly restructured, less focus is placed on the relationship of industry to community and of management and workers to one another. The silent contract is based on the interdependence of workers and managers and, to some extent, the interdependence of community and industry. As work changes, the forms of those relationships may change, but their intrinsic importance remains. Attention to the issues identified in this chapter may be useful in developing organizational structure and culture fitted to our current economy but still connected to values that maintain the continuity of the silent psychological contract.

## REFERENCES

Alderfer, C. P., Alderfer, C., Tucker, L., & Tucker, R. C. (1980). Diagnosing race relations in management. *Journal of Applied Behavioral Science, 16,* 135–166.

Alderfer, C. P., & Klein, E. B. (1978). Affect, leadership and organizational boundaries. *Journal of Personality and Social Systems, 1,* 19–33.

Astrachan, J. H. (1988). Family firm and community culture. *Family Business Review, 1*(2), 165–190.

Astrachan, J. H. (1991). *Mergers, acquisitions, and employee anxiety: A study of separation anxiety in a corporate context.* New York: Praeger.

Astrachan, J. H., & Herbert, J. I. (1988). *An intergroup simulation: The collared exercise.* Unpublished manuscript, Yale University.

Berenbeim, R. F. (1985). *Company programs ease the impact of shutdowns (A research report).* New York: Conference Board.

Harshbarger, D. (1987). Takeover: A tale of loss, change and growth. *Academy of Management Executive, 1*(4), 339–343.

Iglehart, J. K. (1995). Health policy report. Academic medical centers enter the market: The case of Philadelphia. *New England Journal of Medicine, 333,* 1019–1023.

Kassvier, J. P. (1995). Managed care and morality of the marketplace. *New England Journal of Medicine, 333,* 50–52.

Leigh, P. (1994, October 3). Specialists face the future: Early signs of a shakeout. *American Medical News, 37*:1.

McGregor, D. (1960). *The human side of enterprise.* New York: McGraw Hill.

Rivo, M. L., Mays, H. L., Katzoff, J., & Kindig, D. A. (1995). Special communication: Implications for the physician workforce and medical education. *Journal of the American Medical Association, 274,* 712–715.

Rosenberg, C. E. (1987). *The care of strangers: The rise of America's hospital system.* New York: Basic Books.

Samuelson, R. J. (1995). Our great economic lesson. *Newsweek* (May 22), p. 67.
Schaar, J. (1970). Legitimacy in the modern state. In P. Green & S. Levinson (Eds.), *Power and community: Dissenting essays in political science* (pp. 276–327). New York: Pantheon Books.
Starr, P. (1982). *The societal transformation of American medicine.* New York: Basic Books.
Whitcomb, M. E. (1995). Sounding board: Correcting the oversupply of specialists by limiting residencies for graduates of foreign medical schools. *New England Journal of Medicine, 333,* 454–456.

# 3

# The Tiller of Authority in a Sea of Diversity:

## Empowerment, Disempowerment, and the Politics of Identity

*Geoffrey M. Reed, Ph.D., and
Debra A. Noumair, Ed.D.*

"Diversity" has become a catchword of the 1990s, a framework that implies that a wide variety of differences among us deserve recognition, respect, and protection. Such a framework appears to promise a more complex, integrative model of group relations and individual identity. For example, a brochure for a recent conference focusing on themes related to diversity, cosponsored by the Washington-Baltimore Center of the A. K. Rice Institute and the Office of Continuing Studies at American University, asserts that "no longer can the struggle be reduced to such basic dichotomies as black or white, male or female, rich or poor, old or young, American or international, gay or straight, oppressor or oppressed." Our political leaders frequently reinterpret such differences as valuable and indispensable to a richer whole, with metaphoric references to tapestries and rainbows. According to R. Roosevelt Thomas, Jr., of the American Institute for Managing Diversity, "If you have a jar of red jelly beans and add some pink and green ones, diversity is not the pink and green jelly beans, but the resulting mixture" (MacDonald, 1993, p. 25).

At the same time, the grimness, seriousness, and violence of the conflicts that underlie the issue of "diversity" have never been more apparent. The entire nation watched as all of Los Angeles seemed to be in flames in response to the verdict of an all-White jury that appeared to convey the message that White policemen could brutalize or even kill Black men with impunity. There were diametrically opposed reactions to the recent trials of several African American men charged with the attempted murder of a White truck driver during these disturbances. Some viewed these men as scapegoated by a judicial system implacably hostile and fundamentally unfair to Blacks, and cast them as political prisoners by referring to them as the "L.A. Four." For these people, the verdicts of innocent on most of the charges were cause for jubilation. Others viewed with dismay and fear the apparent fact that innocent people could be beaten and severely injured, simply because they were White, without swift and severe punishment. The deep differences in the worldviews of many Black and White Americans have never been so apparent or so public as in the aftermath of the O. J. Simpson verdict. Turning to issues of gender and sexual orientation, we note that terrorism and even murder have become commonplace in the battle over the reproductive rights of women. In towns, cities, and states across the country, far-right groups organize movements to make it illegal to offer protection from employment discrimination to gays and lesbians, and hate crimes against gays and lesbians have continued to increase. Recently, a Virginia court took a mother's children away from her, ruling that the mere fact of a woman's lesbian relationship rendered her an unfit mother, regardless of the nature of her relationship with her children or the environment she provided.

One of the risks of the semantics of diversity is that these deeper and unresolved problems may be obscured by the new rhetoric of the bean jar as much as by the old rhetoric of the melting pot. The cover of diversity may be used to avoid some of the most fundamental conflicts we face as a society. For example, it is difficult to accurately identify and discuss the fundamental problems of American racism when these issues are couched in terms of diversity. African Americans become simply one of many politically defined subgroups competing on ostensibly equal terms for attention and resources, as do Native Americans and others who carry long and deep histories of racist oppression. Thus, devastating injustice may be trivialized through discussions of diversity. Diversity can be "celebrated" or "managed," while racism, sexism, and homophobia cannot. Further, the semantics of diversity may be used specifically to exclude consideration of particular issues and problems. For example, the current, bitter struggles over the treatment and rights of gay and lesbian citizens are explicitly excluded from consideration in discussions of "cultural diversity," "ethnic diversity," or "multiculturalism."

Since we point out that semantics frame the political issues, perception, and discussion of diversity, a comment about our own terminology in this chapter seems appropriate. Specific subgroups that are most identifiable on the field of diversity are most frequently referred to as minorities. However, the term "minority" implies a one-down relationship to the "majority." This term reinforces the specific social perception that "minority groups" are distinct constituencies comprised of small numbers of individuals who demand special rights and special treatment at the expense of most other people. For example, movements to prevent the extension to gay and lesbian citizens of basic civil rights granted without question to others have achieved particular success by casting gays and lesbians as demanding "special rights." The term "minority" also defends against the fear that if these groups united they might constitute an actual majority, and that American Whites may be outnumbered by members of non-White ethnic groups in the foreseeable future. However, there is not an equivalent term to denote these groups. Therefore, we have used the terms *minorities* and *minority groups* at times throughout this chapter, but have often placed them in quotation marks to highlight these dynamics. There is also some social debate over the use of the terms *African American* and *Black*. Clearly, not all individuals who are socially identified as "Black" are of ancestry that can be most immediately traced to Africa. However, the connotations and political implications of the term *Black* have been justifiably questioned. In this chapter, we consider issues related to race, particularly to "Africanism" (Morrison, 1992). However, we have to some extent used the terms *African American* and *Black* interchangeably, although with some awareness of their different implications and connotations.

In addition to these broader social themes, issues of "diversity" are currently prominent within organizational, institutional, and industrial contexts. In an article entitled "The Diversity Industry," MacDonald (1993) cites estimates that 40% of American companies have instituted some form of diversity training, and that half of the Fortune 500 companies have someone within the organization with responsibility for diversity. This figure grows each year. She describes the burgeoning field of diversity consulting, with practitioners charging companies average fees of $2,000 per day. Similarly, universities have struggled with demands for greater numbers of, more powerful, and increasingly specific cultural studies departments. There has been some thoughtful concern that racial separatism and infighting between groups on campus is fueled by this trend, as well as more predictable, less thoughtful backlash by White students, administrators, and faculty. Amid these struggles, the political use of the "diversity" trend to increase the power of various subgroups is clear. Recent protests at the University of California, Los Angeles (UCLA) focused on the creation of a Latino studies *department,* in spite of the fact that there was an existing interdepartmental program. Thus, the goal of the protest seems not to have been to make Latino studies courses available to students, since these already existed, but to gain greater power and autonomy for this program within the university. Parity with African American studies programs was also raised as an issue.

Our goal in this chapter is not to examine whether diversity management programs or cultural studies departments are necessary or appropriate, or to determine what form they should take. Rather, our goal is to articulate and clarify the political nature of issues related to diversity, and the obstacles and resistance to meaningful social discourse, learning, and change related to these themes. These dynamics are generally unspoken and covert, and frequently unconscious. We use both our personal perspectives and the theoretical framework of group relations as a point of departure. We argue that when we undertake discussions about diversity—as leaders, consultants, group members, or as a society—we must acknowledge the political nature of these discussions. Our goals and agendas when we take up these issues must be clear. There is an underlying assumption that examination of these issues will lead to social learning that will somehow be valuable and helpful. This vague hope is not sufficient. True social learning leads to change. If our goal or our effect is to produce change, we need to be very clear about what we have been authorized to do by whom—hence, holding the tiller of authority in a sea of diversity.

Further, as we develop in this chapter, many of the dynamics surrounding diversity function to preserve the status quo, rather than to support change of any kind. We must be conscious of the risk of being co-opted by these forces. We must be clear about how our work may be used to justify and shore up existing structures and institutions, while creating the appearance of addressing issues of "diversity." Specifically, conferences or task forces on diversity may be used by organizations and institutions as tools of oppression. Without meaningful authorization and commitment to structural change, these efforts may be no more than window dressing, serving to mollify discontented members of oppressed subgroups and to quiet the disturbances they create.

Much of our perspective in this chapter comes from a series of conferences on diversity sponsored by the New York Center of the A. K. Rice Institute, and conducted at Teachers College at Columbia University from 1992 through 1995. The first of these conferences had been requested in large part due to an institutional disturbance at Teachers College created by gay and lesbian students. During a case conference, a student (not gay or lesbian) had presented the case of a homosexual man she was seeing in psychotherapy. She indicated that both her supervisor and the professor of her course on object relations viewed homosexuality as inherently psychopathological. She said that this was confusing for her because this was not her experience in treatment with this man. Instead, his homosexuality seemed to her to be simply a part of his identity and situation, and not something that in itself should be a focus of treatment. The views of her supervisor and professor were not countered by any of the faculty members present at the case conference. The subsequent protest of these events by gay and lesbian students eventually led to the creation of a "task force" on diversity (led by a junior faculty member), and to the request for a conference.

No one with significant institutional authority attended the conference. Only junior and adjunct faculty members, who did not have the power to create meaningful changes in the institution, were present. We use this example, and will return to it later, to highlight the risk that such task forces or conferences on diversity may function primarily to create the appearance of addressing "minority" concerns, preventing these subgroup members (in this case gay and lesbian students) from demanding more basic, profound, and real change in an institution that continues to exist long after the temporary institution of a conference has ended.

## OUR PERSPECTIVE

We believe that it is important for us to articulate our personal point(s) of departure for the remarks that follow. We met, quite literally, at the staff–group member boundary at the 1992 Diversity and Authority Conference at Teachers College, Columbia University. We use incidents from that conference, and several others focusing on these themes, throughout this chapter to illuminate our discussion. In addition, it is clear that our own perspectives on diversity are framed by the lenses of experience and identity through which we view the world. We wish to acknowledge our similar, yet different, personal frames of reference.

The first author enters this discourse as a White, gay man, who has had access to the privileges of being White and male, but who has also suffered personal and institutional discrimination because of his homosexuality. Containing both "oppressor" and "oppressed" aspects of identity contributes to his perspective in this chapter. The second author is a woman who identifies as White, but whose skin is darker than that of most Latinos and many African Americans, and who is often perceived by others as non-White. This response to the color of her skin and her features deeply influences her identity and results in her experience of being on the boundary between Whites and people of color. For both of us, "passing" in dominant society has afforded privilege, and at the same time contributed to the experience of being on the margin, at risk for being found out and rejected based on other aspects of who we are. In a sense, then, we both enter this discourse on diversity at the boundary, containing opposing positions within our own identities.

The fact that collaboration across these boundaries is complex and difficult has been obvious throughout our work together. In order to collaborate at the staff-member boundary when we met, we had to overcome the shame of not knowing, the one-down position of a less powerful group in relation to a more powerful one. We have struggled with similar feelings, particularly the perceived necessity for one light to be dim so that another may be bright, throughout our work on this chapter. Indeed, the academic model of authorship has no mechanism to identify equal collaboration, but is a model of dominance, the ranking of one person's contribution over another's. We wish to acknowledge our differentiated, interdependent model of working together, and our equally important contributions to this chapter.

## THEORETICAL ORIENTATION

Originally developed at the Tavistock Institute for Human Relations, the theoretical roots of the Group Relations model can be traced to the work of Wilfred Bion and Melanie Klein (see Rioch, 1979). The group relations perspective highlights such issues as power, authority, and leadership in group and institutional life. The group relations model assumes that group process always occurs on two levels, overt and covert (see Rice, 1965). The overt refers to what is explicit, the content of what is being communicated, the group's stated purpose, tasks, and self-conception. The covert refers to what is taking place under the surface, what is implied and not stated, the dynamics and process of group functioning, and is often unconscious. A premise of the Group Relations model is that in the same way that infants struggle with conflicting wishes to merge and separate from the mother, who has the power to both gratify and frustrate basic human needs, group members struggle, generally unconsciously, with conflicting desires to join and to remain separate (see Wells, 1990). In order to tolerate the tension between engulfment and estrangement, to manage the anxiety and ambivalence that is inherent in group life, defense mechanisms are employed. These mechanisms are fundamental methods of simplifying the contradiction, ambiguity, paradox, and multiple realities inherent in human experience.

Splitting and projective identification are two of the most commonly used defenses. At its most basic level, splitting refers to the process of dividing the world (and others) into all good and all bad. Splitting also describes the polarization of specific characteristics as contained within individuals or subgroups, whereby all of a particular quality is perceived as being contained within one, and all of its opposite contained within another. Projective identification refers to the process by which individuals and groups deposit undesirable or ambivalently held aspects of the self and/or group into other individuals or groups, and is a mechanism through which splitting is accomplished when it involves one's own individual or group identity. The individuals or groups receiving these projections are subtly and often unconsciously encouraged to behave in ways that are consistent with these projections, since they are expressing these aspects on *behalf* of those who projected them (i.e., through identification). Individuals or groups who are chosen to express others' feeling do so because they have an unconscious disposition, or "valence" for such expression, and/or because of demographic characteristics. To the extent that group members involve themselves in disowning unwanted aspects of themselves, and taking on unwanted aspects of others or of the group, they tacitly contract to participate in an interdependent, collusive process. It is precisely through this process that group members are connected to one another. This unconscious agreement constitutes the covert level of group life and creates the group-as-a-whole. Individuals and subgroups represent not only themselves, but also some part of the larger group or system.

## SOME SOCIAL DYNAMICS OF IDENTITY

Group relations methodology allows us to examine the ways in which groups, organizations, and institutions use subgroups and individuals as spokespersons, leaders, scapegoats, heroes, or enemies because of stereotypical assumptions about those subgroups based on such variables as race, gender, and sexual orientation. We view the defenses of splitting and projective identification as powerfully related to the issues surrounding "diversity," particularly to those that are most explosive and divisive and create the greatest obstacles to collaboration across group lines.

Both individuals and groups actively construct self-identities from the available set of roles through processes of internalization and identification. Externalization and rejection of other potential identities, then, is the corollary of this process of identity construction. A fundamental part of both the individual and intragroup defense systems is the use of the "other" as a container for undesirable aspects of the self (Bion, 1959; see also Skolnick & Green, 1993). This is accomplished in large part through splitting and projective identification. Aspects of the self or the group that are disowned or rejected are projected onto others, so that desirable characteristics are contained entirely within the self or one's own group, and their undesirable counterparts in the "other." At the group level, this process is an important basis for stereotypes. Thus, projections and stereotypes are among the building blocks from which self-concepts, public personae, and group identity are constructed. The "not me" and "not us" are used to define "me" and "us."

In her book of literary criticism entitled *Playing in the Dark: Whiteness and the Literary Imagination,* Toni Morrison (1992) describes the ways in which African Americans have been used throughout American history to contain disowned aspects of dominant White society, a process she refers to as Africanism. "Africanism is the vehicle by which the American self knows itself as not enslaved, but free; not repulsive, but desirable; not helpless, but licensed and powerful; not history-less, but historical; not damned, but innocent; not a blind accident of evolution, but a progressive fulfillment of destiny" (p. 52). Morrison's central thesis is that for White Americans, the ideals and experiences of freedom, individualism, manhood, and innocence have depended on the existence of a Black population that is manifestly *unfree,* and which serves Whites as the embodiment of their own fears and disowned desires. Women have been used in a similar way to contain projections of being weak, passive, and powerless, as well as being seductive, manipulative, devouring, and castrating. Gays and lesbians have carried disowned projections of sexual depravity based on the fear that unbridled expression of sexual impulses will lead to the collapse of the institutions of family and society. Of course, it is *one's own* sexual impulses, not those of others, that are the real source of terror. It has been argued that sexism and homophobia are linked through the use of femaleness and homosexuality (particularly male homosexuality) as "not me" to define American male identity.

A related, but more subtle, phenomenon is that victimization is a source of identity for oppressed groups, who are unified and emotionally connected by the oppression that they have experienced. It is well established in the social psychological literature that external threat is one of the most powerful factors increasing group cohesion (e.g., see Sherif & Sherif, 1965). It is difficult to imagine the form that modern African American identity would take if slavery had never existed, and if no hierarchy of freedom, power, and desirability had ever been based on skin color. As a more specific example, sexual historians point out that while homosexual *behavior* has apparently existed throughout human history, homosexuality as an *identity*, as a *type* of person, is a relatively modern phenomenon. The crystallization of homosexuality as an identity may be traced, in part, to the development of social ideology directed toward oppressing homosexual behavior. Powerful social movements focusing on "vices" during the late 19th century Victorian era, as well as the systematic witch hunts and purges conducted in America from the late 1940s to the early 1960s against homosexuality along with Communism were among the factors that led to the evolution of homosexual communities, homosexual lifestyles, and individuals who defined themselves specifically in these terms (e.g., see D'Emilio, 1983; Rubin, 1984).

## THE POLITICAL USE OF IDENTITY

As may be inferred from the preceding examples, it is our position that at the heart of discussions regarding "diversity" are issues related to power. Our use of the term *power* here is intended to encompass a range of meanings. We include access to and control of material, social, and institutional resources. We include meaningful enfranchisement in political and organizational contexts, and the ability to shape and influence decisions, agendas, and priorities. We include the ability to act instrumentally on one's own behalf or in the service of one's own interests. We include connection with a network of others with similar interests, when acting in concert with these others produces greater results than acting in isolation. Perhaps most importantly, we include the power *to define* social reality, experience, and history (Sampson, 1993). However, many of the dynamics that connect issues of diversity with issues of power are generally unspoken. We attempt to articulate some of them here.

A core cultural assumption about power is that it is inevitably scarce. We view our world as one of limited resources. We maintain the fiction that some must go hungry because there is not enough food, some must be poor because there is not enough money, some must be homeless because there is not enough shelter. The combination of the assumption of scarcity and the American "myth of meritocracy" (McIntosh, 1992) allows us to maintain the belief that people deserve what they get and get what they deserve, justifying the inequities of our system. Similarly, we act as if there is a sharply limited quantity of political and institutional power, access, and influence. By necessity, a few will have this power, and most will not. The assumption that scarcity is the necessary state of the world and the necessary model for power relations masks the political choices we have made as a society. The assumption of scarcity shapes social discourse on diversity in fundamental ways, although the underlying questions are rarely raised explicitly. For example, if only some will have power, if only a few will have access to resources, who will this be? Which groups are deserving, and which are not? The assumption of scarcity suggests that the best bet for each group and each individual may be to try to obtain as much as possible for oneself, regardless of the consequences to others.

The political nature of identity becomes apparent when individuals are pressed to identify themselves as belonging to one group or another. In these situations, such identification frequently has direct consequences for the power base of the various groups involved. It must be acknowledged that presenting a unified front has been a historical necessity for the survival and welfare of many oppressed groups. The civil rights and organized labor movements are obvious examples. However, our point here is that the choice of public identities is in large part determined by this process of "constituency building" (Alderfer, 1983). This leads to strong pressure on individuals to emphasize specific identity characteristics at the expense of others, depending on which characteristics are most politically valuable within a particular setting.

Within institutional settings, questions of diversity are frequently raised in relation to the allocation of resources—programs, positions, time, funding, and so on. If a group is perceived as systematically disadvantaged, then a case can be made that this group is deserving of proportionately greater institutional resources. There may be significant intergroup conflict related to which groups are most deserving of corrective advantage. Identity groups and individuals may minimize their own advantage or emphasize their disadvantage within the context of such discussions. We refer to this as the relationship between context and currency: What chips are worth the most in what context, and how is public identity selected and displayed to others on this basis?

Historically, perhaps the defining characteristic of "minority" identity is the lack of access to positions of power and influence. For example, South African Whites carried the "majority" identity for many years because of their control of resources and political power, in spite of actually comprising a small minority. Discussions of themes related to diversity often focus on demands made by various subgroups to be included in the existing power structure, often without otherwise changing it. Thus, these discussions are based on and framed by the existing model of power relations and the assumption of scarcity (what Sampson, 1993, refers to an monologuism), rather than aimed at the conceptualization and development of an alternative model. Within this context, representation becomes a central issue. The assumption is that if individuals who share the externally identifying characteristics of oppressed subgroups are among those at the inner circles of power, this will improve conditions for other members of that group.

However, this dynamic often leads to reductionism, and contributes to the failure to address the real, underlying issues. For example, the Los Angeles Unified School District recently needed to select a new superintendent. The number two position with the district had been occupied for some time by a Black man, who seemed to be best qualified to fill the post. A large contingent in the community said that because of the high percentage of Latino students in the district, a Latino candidate must be appointed to the position. The Black candidate's record and sensitivity in responding to the needs and concerns of Latino students and the Latino community was not generally discussed in the acrimonious debate that followed. Similarly, as a nation we have examined the demographic characteristics of our president's appointments, without reference to their records in relation to the communities they are assumed to represent.

This reductionism creates the opportunity to twist concerns regarding representation to serve policies of racism, sexism, and homophobia. Representatives of "minority" groups who do not share many of the experiences of oppression, or who are actively hostile to the agendas of groups they ostensibly represent, may be anointed by those in power. Unless deeper questions are asked, the existing hierarchy may consolidate and perpetuate its own power while appearing to satisfy explicit demands for representation, creating a double bind for vocal "minority" groups. It is relevant to point out that our only African American Supreme Court justice is Clarence Thomas, and that a lesbian presided over systematic discrimination against homosexual artists as head of the National Endowment for the Arts during the final months of the Bush administration. As long as the assumption of scarcity provides the framework, access and influence will be denied to the many by the few.

## THE MYTH OF REDRESS: A BASIC ASSUMPTION OF DIVERSITY GROUPS

Within groups focusing on the theme of diversity, there is frequently an unconscious fantasy held by the group that appears to share many of the characteristics of basic assumption functioning in groups as described by Bion (1959). Basic assumption functioning is a regressive state in groups that is characterized by specific, generally unconscious beliefs about the purpose and life of the group. In groups focusing on diversity, the operative unconscious belief is frequently that the group has met for the purpose of correcting the injustice of oppression experienced by subgroups collectively and historically, and by group members individually. We refer to this as the myth of redress—that is, that resources and power will be redistributed to correct the injury and oppression that have been experienced by subgroups and by each member.

Basic assumption beliefs are "invested with reality by the force of the emotion attached to them" (Bion, 1959, p. 147). The myth of redress is fueled by the sense of victimization and narcissistic injury that each member—regardless of ethnic/racial status, gender, or sexual orientation—has experienced at the hands of an unjust world. Powerful, primitive emotions of rage and loss are attached to these experiences. These emotions give rise to a sense of entitlement, including entitlement to concrete, material symbols of power. The looting that occurred during the 1992 civil disturbances in Los Angeles may be partly understood in these terms. Although it is widely perceived that the individuals participating in the looting were uniformly Black, this was not the case. A story reported in the *Los Angeles Times* described a group of White women in a Jeep Cherokee seen looting The Gap on Los Angeles's most trendy shopping street. The emotional justification for such actions is provided by a nearly universal sense of victimization and narcissistic injury, based on the framework of scarcity.

As described by Bion (1959), "any leader is ignored by the group when his behavior or characteristics fall outside the limits set by the prevalent basic assumption" (p. 171). When the myth of redress is operative, the consultant and any potential leader are evaluated based on the group's perceptions of the likelihood that he or she will act to compensate for the oppression of subgroups and individual members. The presence or absence of "diverse" characteristics in the person of the consultant is viewed as signifying the extent to which he or she may do this, and the extent to which he or she will be responsive to or allied with the claims of particular subgroups. In one diversity conference, the small group to which the first author was consulting devalued and ignored him, explicitly because of his being a White male. In fact, the first comment made during the life of the small group was the expression by one group member of her disappointment to come to a conference on diversity and find herself assigned to a White male leader. On the second day, the information was offered by another member that the consultant was gay. This information had been held by specific group members, but was concealed until the consultant interpreted this fact rather vigorously. At that point, the group's response to the consultant shifted dramatically, as though he was no longer a White male. The gay and lesbian group members were given considerable authority. A powerful "gay and lesbian alliance" emerged, and the oppression experienced by homosexuals became an important matter for group consideration.

The myth of redress underlies competition among subgroups for "the throne of disadvantage" (Gates, 1993, p. 42). The group belief is that if redress for past oppression will be made, it should be given to subgroups according to the degree of victimization they have experienced historically and personally. Competition may be seen as a struggle for who will be first in line. The nature of this competition is discussed in more detail in a later section. When the myth of redress is operative, oppression is power; victimization is the group "currency." Group members identify themselves to others according to the aspect of their identities that most effectively carries the label of victim. Thus, a middle-class African American woman is likely to identify herself as Black, omitting reference to the advantages of her class. A White man is likely to emphasize his impoverished background. Aspects of identity that may fade to the background in other settings, such as being Jewish or gay, become the most prominent aspects put forward. Personal histories of oppression are frequently recited, both by members and by staff. Medals of victimization signify rank in the diversity hierarchy and access to the "truth" about the relevant issues. In this land of the one-eyed, the blind man is king. As suggested by the group's inability to contain both the gay and White male parts of the first author's identity simultaneously in the preceding example, and by the selection of narrow pieces of identity for presentation to others, the myth of redress frequently leads to fragmentation. Individuals are reduced by the group to a sum of pieces through which they can lay claim to the experience of victim. According to this logic, "African American woman" has a higher diversity score than "gay White man," but one that is lower than "physically challenged Latina lesbian."

Rather than offering an alternative model of power relations, groups operating under the myth of redress create a mirror image, bound and determined by the external reality it reflects. Standing the existing model of power relations on its head, however satisfying in the short term, does not provide a viable alternative. The myth of redress fails because ultimately, of course, no redress is possible. Within American society, it is not possible to devise a corrective for the horror of slavery, or for other social, institutional, and individual forms of oppression. The pain of these experiences cannot be fixed. This is *not* an argument for dismissing our histories of oppression of and by others. Simply saying there is nothing we can do maintains the status quo at society's peril. In order to move beyond our past, to develop an alternative model of collaboration, we must acknowledge our collective participation in the oppression of others, and experience the sorrow related to our powerlessness to change our history. Only through such a process can we negotiate a mutual reality, a shared—though differentiated—experience (i.e., a dialogue; Sampson, 1993) that can be a starting point for a collaborative model. This involves a great deal of pain *on both sides,* and "minorities" cannot do this alone. The fragments must be integrated into a more complex whole. The fear and pain of doing this, and defensive group functioning based on the myth of redress, present a significant obstacle to the construction of a more integrative model. We turn to this theme next.

## RESISTANCE TO CHANGE

Bion (1977) asserts that significant social learning involves change *that is invariably resisted by the prevailing establishment* (see also Skolnick & Green, 1993). History offers no ready example of a case in which redistribution of power and resources has been willingly granted by those in control of the system, and there is little reason to expect that the current situation will be different. However, to assert that the reluctance of those at the top to relinquish their power is the major form of resistance to change is to collude with the polarizing dynamics that maintain the status quo. Change is terrifying for everyone. As oppressive as the current authority structure may be, the comfort of familiarity is favored over facing the terror involved in change. It is far easier to express anger at the current system than it is to create and take responsibility for a new one.

For oppressed groups no less than for their perceived oppressors, change from the existing model of power relations threatens both identity and community. For example, a new model may mean loss of entitlement and special status, loss of an explanation for failure, and an end to the exemption from facing the victimizer in oneself (a theme we develop in a following section). The abandonment of group defenses seems to be equivalent to annihilation, and gives rise to so much anxiety that change is experienced as catastrophic (Bion, 1977). The existential terror resulting from this threat is likely to be suppressed and maintained at an unconscious level, but is manifested as powerful resistance to change.

In the next sections, we explore in more depth some of the specific manifestations of this resistance, which pose obstacles to a new model of collaboration. Only if these forms of resistance are available to conscious work is there any hope that they may be overcome. We discuss three major forms of resistance. The first is denial of the victimizer role, which serves to trap various subgroups in their function as containers of the experience of oppression. Emergence from this position is impossible as long as the reality of this experience, and the role of others in creating and maintaining it, is not acknowledged. Many of the ideas in this section were developed by Williams (1993) in her paper entitled *The Dilemma of Being Both Victim and Victimizer,* and are repeated here with her permission. The second form of resistance we discuss is fragmentation and competition among subgroups. As long as divisions between various "minorities" are maintained, no effective challenge to the dominant hierarchy (*in fact* a minority) can be mounted (Noumair, Fenichel, & Fleming, 1992). Finally, we discuss how, even when invitations are made to participate in changing the system, the failure to provide meaningful authorization dooms these efforts to failure. These forms of resistance operate together, generally unconsciously, to preserve the status quo and to undermine meaningful collaboration and change.

## Denial of the Victimizer Role

While the mayors of our cities and consultants to our corporations talk about "celebrating diversity," there is general silence about the victimizer role. As discussed, the tendency during discussions about "diversity" is for everyone to identify some piece of identity through which he or she may lay claim to the experience of oppression. In this way, the experiences and history of being "Irish," for example, or even blonde and beautiful, are used to articulate a sense of personal victimization. A claim to the experience of oppression is used to deny participation in the oppression of others, as though victim and victimizer are mutually exclusive positions.

This is one way in which those who enjoy privilege deny their participation in a system which confers dominance at the expense of "minority" groups. As McIntosh (1992) states, "Obliviousness about white advantage is kept strongly inculturated in the U.S. and serves to maintain the myth of meritocracy, the myth that democratic choice is equally available to all" (p. 81). Williams (1993) points out that this statement also applies to male advantage and heterosexual advantage. As a result, the experience of victimization—either as subject or object—is identified with "minorities," and the moral and ethical dilemmas of privilege remain unconscious and unexamined. Experiences related to oppression are kept within "minority" group members. African Americans, women, gay men, and lesbians are expected to teach members of dominant groups about "diversity," and, within organizations and institutions, to carry out the work related to this issue. At the same time that members of disenfranchised groups are expected to become advocates for the less powerful, they are frequently accused of seeing only a limited piece of the picture, thereby sabotaging the very work that is expected of them. This perpetuates the denial by members of privileged groups that they have any direct access to or experience of these issues, allowing them to remain unconscious of their role. The need for shared responsibility is obscured (Fouad & Carter, 1992), undermining the possibility of change. Thus, leaving this work to "minorities" guarantees that the status quo will be maintained and left unexamined.

Denial of the victimizer role is apparent in conferences about diversity. For example, during an early small study group session of a recent conference on diversity, an attractive, young, Southern, White woman described her pain and outrage at having been described as "perky" during a recent job interview. The sexism of this remark is clear and offensive, and there is no doubt that her feelings about it were authentic. Within the group, however, her description of this experience functioned to deny the possibility that she, as a victim of oppression herself, could participate in the victimization of others. This may be understood as a preemptive (and successful) effort to ward off anger and rage that might be directed at members of dominant groups by "minority" members within the context of a conference on diversity. Her equal claim to the experience of oppression was unchallenged by other group members throughout the conference. Through such defensive maneuvers, the dynamics of oppression are made unavailable for group work, since the oppressor is "out there" and "back then," disavowed by and disconnected from any of the people in the room.

It must be pointed out here that the victimizer role is also denied by members of "minority" groups, and that their status as oppressed often provides exemption from examining these issues within themselves. The reality is more complex and painful. Throughout their history of interaction with dominant White society, Blacks have participated in the oppression of other Blacks, just as women have oppressed other women, and gays (such as Roy Cohn) and lesbians have oppressed other gays and lesbians. Alice Walker (1992) begins her book *Possessing the Secret of Joy*, in large part an examination of this theme, with a quotation from a bumper sticker: "When the axe came into the forest the trees said the handle is one of us."

This is partly related to identification with the aggressor, or internalized oppression, but is amplified by the difficulty in maintaining multiple aspects of identity simultaneously: that is, victim *and* victimizer. Individuals who are members of oppressed groups along one dimension may be members of majority groups along others. Thus, Black men may victimize Black women and straight Black women may perpetrate heterosexual oppression against gay and lesbian people at the same time that they are themselves victims of racial or gender oppression. However, Williams (1993) points out that for Blacks to acknowledge the role of victimizer is to reinforce stereotypes of being dangerous, violent, murderous, and predatory. "This may also be true of women who are stereotyped as manipulative and devouring, and gays and lesbians whose very existence is considered a threat to children, public health, and so-called family values" (Williams, 1993). Thus, a double bind is created. Members of oppressed groups cannot be expected to speak to their own participation in the victimization of others unless Whites, men, and heterosexuals do so first.

## Fragmentation and Competition among Subgroups: Divide and Conquer

An important implication of the assumption of scarcity is the belief that the concerns and demands of one subgroup may be heard only at the expense of others. "Minority" groups are pitted against another, preventing any unified demand for change. Subgroups generally accept the framework of scarcity without question, and demand to be included at the table of power, rather than demanding structural change in the way that power is conceptualized and allocated. Direct competition between subgroups is based on and supports the existing framework, because of the implicit assumption that only one place can be made at the table. For example, prior to the first diversity conference at Teachers College, a meeting was called to discuss the concerns of gay and lesbian students that had been raised following the student presentation described at the beginning of this chapter. During this meeting, a coalition of students of color presented a prepared statement protesting the institution's attention to issues of importance to gay and lesbian students when other issues of concern to students of color had not been addressed. After presenting the statement, these students left the meeting, closing off any opportunity for collaboration and any possibility of more unified calls for change in the institution.

This example highlights a particular tendency for African Americans to see other groups as interlopers in their struggle for equality. These other groups are seen as appropriating and copying the civil rights struggle. Cultural historians point out that this is an oversimplification, and that distinct social dynamics led to the civil rights movement, the women's movement, and the struggle for gay and lesbian rights (e.g., Harris, 1981). From the African American perspective, however, these other struggles create the risk that the fundamental problem of racism in American society will be ignored and their oppression will continue amid the clamor of other groups. In his thoughtful essay on this topic, Henry Louis Gates, Jr., (1993) quotes a prominent civil rights veteran, who on the day of the recent gay and lesbian March on Washington wrote that "It is a misappropriation for members of the gay leadership to identify the ... march ... with the Rev. Dr. Martin Luther King Jr.'s 1963 mobilization" (p. 42). Gates goes on to describe the surprisingly widespread sentiment that "gays are pretenders to the throne of disadvantage that properly belongs to black Americans, that their relation to the rhetoric of civil rights is one of unearned opportunism" (p. 42).

Gates asserts that this point of view can only be held by ignoring the existence of Blacks *who are also gay*. He describes the example of Bayard Rustin, a Black gay man who was the organizer of the 1963 civil rights march on Washington. Rustin was prevented from being named the director of the march because of his homosexuality, and the fear that it would be used to discredit the mobilization. The title went to A. Philip Randolph, who accepted it on the condition that he could deputize Rustin to do the actual work of coordinating the mass protest.

This fragmentation maintains divisions between subgroups, rendering parts of the self unavailable. Projections cannot be reintegrated, a more complex whole cannot be acknowledged and held, and such conflicts are unresolvable as long as these divisions are maintained. These dynamics can also be seen in the tensions between Blacks and Jews, or the violence between African American community members and Korean owners of convenience stores during the 1992 civil disturbance in Los Angeles. Collaboration is prevented, ensuring that there will never be enough united opposition to the power structure to mount a successful rebellion. Subgroups are maintained as "minorities" rather than united as a new majority. Thus, this competition represents an unconscious collusion with the status quo.

As discussed, the competition for the yellow star of oppression, related to whose pain and anger is more profound and whose agenda and needs are more deserving, emerges vigorously in conferences on diversity. Significantly, Whites are generally silent while this subgroup competition plays out, except to offer their own experiences of oppression in order to prevent their identification as victimizers. On the surface, it would appear that this silence is related to the fact that in an environment that attaches a premium to diversity, they have no chips to play. On the other hand, it is clear that as long as competition among subgroups continues, there can be no real threat to the established order or to their primacy within it. The competition serves to detour the greater and more dangerous conflict with members of privileged groups. The competition may also be exploited more actively, by appearing to accede to the demands of the most powerful or vocal subgroup, thus silencing this group or separating it from others with whom an alliance might be formed. "Diversity" is used as a cover for the preservation of power. We have already mentioned George Bush's nomination of Clarence Thomas for Supreme Court justice, and appointment of Anne-Imelda Radice as head of the National Endowment for the Arts as examples of this phenomenon.

## Failure to Provide Meaningful Authorization to Effect Change

A third form of resistance to change is that individuals or groups given responsibility for change are frequently given insufficient authorization to carry out this task. Early in this chapter, we described the response of Teachers College to the concerns of homosexual students as an example of this phenomenon. Within institutions, the appearance of addressing issues related to diversity may be created by designating certain individuals, generally members of "minority" groups, as responsible for these concerns without giving them real power to influence the system. The experience of many individuals on institutional "diversity task forces" reflects this dynamic. In such circumstances, proposals for real change in the power structure are likely to be at best ignored and at worst met with open hostility. Bill Clinton's nomination of Lani Guinier to a top civil rights post was retracted following public and political reactions to her ideas about how greater power and enfranchisement in the political system might be given to minority voters. Her explicitly speculative proposals were painted as a direct assault on the very principles of a democratic society.

The initial sense of gratification and narcissistic inflation that often occurs when members of "minority groups" are asked to represent issues that have previously been kept invisible or unresolved may prevent accurate perception of the absence of real authorization behind this invitation. For example, the first author was flattered by the invitation to be on staff at the 1992 conference at Teachers College, although he was aware that he had been asked specifically because of his homosexuality. In accepting the invitation, he looked forward to the opportunity to address issues related to sexual orientation with integrity and seriousness. Initially, he missed the way in which he was being used by the institution (and the conference) to *contain* the issue of homosexuality by creating for gay and lesbian students the illusion of meaningful action. An institutional commitment to engage authentically in work related to the issue of homosexuality was absent. When this dynamic became apparent in the conference, he experienced his powerlessness to effect any real change, and the demoralization and anger that accompanied that position.

Again, however, to characterize this form of resistance as solely located among those in power is to collude with the polarizing dynamics that maintain the status quo. Frequently, the most bitter attacks on those who wish to change the system come from those on whose behalf these individuals are attempting to act. The history of grass roots advocacy and civil rights organizations is almost invariably characterized by revolving leadership, as each new leader is unable to withstand both the resistance from above and the unanticipated assaults from below. Skolnick and Green (1993) suggest that flaws in the authorization of international organizations such as the United Nations, the World Court, and Amnesty International often cause them to fail in times of crisis, when they are most needed. We suggest that the authorization is insufficient both from those nations who have the resources to back up these efforts, and from those nations who would be most likely to benefit from them.

## TOWARD A NEW MODEL

While diversity is often discussed as a concept that will allow differences to co-exist peacefully in a pluralistic society, we have presented an argument that it is often used as a cover story for the real dilemma, which is the distribution of social power. It is not surprising that interest in diversity has increased at the same time that the redistribution of power has become a major theme in American society and in global politics. Examples of this are plentiful—tax reform, Medicare, Medicaid, welfare reform, term limits, international trade agreements, the fragmentation and bitter conflict in the former Soviet bloc, and others. The frightening question at the core of the most bitter conflicts is "If power is redistributed, who will have it?"

The preceding argument brings us to the necessity of developing a new model of power relations. It is beyond the scope of both this chapter and our ability to create such a model here. However, we believe that this must be a model of differentiated collaboration, arising out of the acknowledgment of our individual and collective participation in the dynamics of oppression, and the negotiation of shared, yet differentiated, reality, experience, and history. Without this negotiation, we are a living Tower of Babel, with no common language and no means of understanding one another (Skolnick & Green, 1993). Developing a collaborative model means addressing fundamental and terrifying questions. If we cannot arrive at this model simply by turning the existing structure on its head, or by admitting selected groups while leaving others without sufficient political voice out, what will it be? Who will gain and who will lose? Who is allowed to propose new models?

We have argued that the changes involved in creating a new model are experienced as catastrophic and as threatening annihilation. As a result, these changes are powerfully resisted, and we have described several specific forms of this resistance. Importantly, such changes are resisted not only and not even primarily by those currently in power. Human history and the limited supply of the world's resources provide powerful reasons to suspect that the struggle for power and dominance of one people by another is inescapable. Ultimately, developing a new model of power relations may be impossible. However, there may be no alternative if we are to meet the challenges that face the world, or possibly even to survive.

## IMPLICATIONS FOR GROUP RELATIONS WORK ON DIVERSITY

The discussion of diversity in groups and organizations, as well as in society at large, necessarily includes attention to power and the allocation of resources. As Noumair, Fenichel, and Fleming (1992) point out, group relations theorists have tended to assume that struggles for power, authority, and leadership can be addressed through the successful reclaiming of projections and the reintegration of disruptive splits (e.g., Skolnick & Green, 1993). Similarly, Fanon (1952/1967) proposes that the polarizing dynamic of oppressor and oppressed can be broken when the oppressed reject the role of abused container without rejecting the humanity of the oppressor. However, the conditions under which subgroups can take up such responsibility and succeed have not been articulated. Is it possible to shift the one-up, one-down structure of group relations, to integrate masculinity and femininity, Black and White, heterosexual and homosexual? Would such integration necessitate the submersion of subgroup identities for the sake of the greater whole?

We must consider seriously the goals of diversity work, rather than assuming that we can be helpful in some unspecified way. Are there goals that can be accomplished? Who has set them and agreed to them? Who is authorized to work in the service of these goals and how are they authorized to work? Is our goal social change? Or is our goal psychological: to be able to see both the oppressor and the oppressed within oneself? If our goal is psychological, what are the social consequences of our work? As leaders and consultants, we need to be sure that we are not unconsciously serving the existing authority structure, or alternatively, using members or those we lead to act out our own revolutionary fantasies, personal agendas, or ideas of desirable social change.

Whatever our answers to these questions, we can expect that authentic work on diversity and authority will be challenging, will take a long time, will be a process, and is likely to involve change that will be experienced as catastrophic and will be strongly resisted. We cannot expect that we can simply import our existing technology to this new set of issues. We will need to develop new models of learning and of being in order to address these issues. If we accept this challenge, we must contain the overwhelming anxiety and terror involved in creating something new—of not knowing the solution, or even if one is possible. This may be the most useful focus for consultation and leadership.

## ACKNOWLEDGMENTS

The authors wish to thank Medria Williams, Ph.D., for her permission to include some of her ideas in this chapter, and for her comments on an earlier version of this manuscript. We also wish to thank Kenneth Eisold, Ph.D., for his helpful comments and encouragement.

## REFERENCES

Alderfer, C. (1983). An intergroup perspective on group dynamics. In J. Lorsch (Ed.), *Handbook of organizational behavior* (pp. 190–222). Englewood Cliffs, NJ: Prentice-Hall.

Bion, W. R. (1959). *Experiences in groups.* New York: Basic Books.
Bion, W. R. (1977). *Seven servants.* New York: Aronson.
D'Emilio, J. (1983). *Sexual politics, sexual communities: The making of a homosexual minority in the United States, 1940–1970.* Chicago: University of Chicago Press.
Fanon, F. (1967). *Black skin, white masks* (C. L. Marmann, Trans.). New York: Grove Press. (Original work published 1952)
Fouad, N., & Carter, R. T. (1992). Gender and racial issues for new counseling psychologists in academia. *Counseling Psychologist, 20,* 123–140.
Gates, H. L., Jr. (1993, May 17). Blacklash? *The New Yorker, 71,* pp. 42–44.
Harris, M. (1981). *America now.* New York: Simon & Schuster.
MacDonald, H. (1993, July 5). The diversity industry. *The New Republic, 209,* pp. 22–25.
McIntosh, P. (1992). White privilege and male privilege: A personal account of coming to correspondence through work in women's studies. In M. L. Andersen & P. H. Collins (Eds.), *Race, class, and gender: An anthology* (pp. 70–81). Belmont, CA: Wadsworth.
Morrison, T. (1992). *Playing in the dark: Whiteness and the literary imagination.* Cambridge, MA: Harvard University Press.
Noumair, D. A., Fenichel, A., & Fleming, J. L. (1992). Clarence Thomas, Anita Hill, and us: A group relations perspective. *Journal of Applied Behavioral Science, 28,* 377–387.
Rice, A. K. (1965). *Learning for leadership: Interpersonal and intergroup relations.* London: Tavistock.
Rioch, M. (1979). The A. K. Rice Group Relations Conferences as a reflection of society. In G. Lawrence (Ed.), *Exploring individual and organizational boundaries* (pp. 53–68). London: Wiley.
Rubin, G. (1984). Thinking sex: Notes for a radical theory of the politics of sexuality. In C. S. Vance (Ed.), *Pleasure and danger: Exploring female sexuality* (pp. 267–319). Boston, MA: Routledge & Kegan Paul.
Sampson, E. E. (1993). Identity politics: Challenges to psychology's understanding. *American Psychologist, 48,* 1219–1230.
Sherif, C. W., & Sherif, M. (1965). Research on intergroup relations. In O. Klineberg & R. Christie (Eds.), *Perspectives in social psychology* (pp. 153–177). New York: Holt, Rinehart & Winston.
Skolnick, M. R., & Green, Z. (1993). Diversity, group relations and the denigrated other. In S. Cytrynbaum & S. Lee (Eds.), *Transformations in global and organizational systems: Changing boundaries in the 90's. Proceedings of the Tenth Scientific Meeting of the A. K. Rice Institute.* Jupiter, FL: A. K. Rice Institute.

Walker, A. (1992). *Possessing the secret of joy*. New York: Harcourt, Brace, Janovich.

Wells, L. (1990). The group as a whole: A systemic socioanalytic perspective on interpersonal and group relations. In J. Gillette & M. McCollom (Eds.), *Groups in context: A new perspective on group dynamics* (pp. 49–85). Reading, MA: Addison-Wesley.

Williams, M. (1993, May). The dilemma of being both victim and victimizer. In Z. Schachtel & D. A. Noumair (Chairs), *The challenges of thematic conferences: Learning from group relations conferences on diversity and authority*. Symposium presented at the 11th Scientific Meeting of the A. K. Rice Institute, Marina del Rey, CA.

# 4

# A Salt-and-Pepper Consultation: Racial Considerations

*Beverly L. Malone, Ph.D., R.N., F.A.A.N.*

The use of multiracial consultancy teams is frequently viewed as positive. The team symbolically represents the diversity of humankind as well as differences that phenotypically are visible and unavoidable. In addition to the symbolism, racial differences in a consultation team highlight the consultants' ability to work together across their differences, visible and invisible. However, there are situations in which positively perceived tactics, such as multiracial consultation teams, are strategically used to mask the unresolved work group issues of organizations and their leaders. The following consultation is an example of this dynamic.

The setting for this consultation was a small university town with a regional medical center located in the deep South. Two female consultants, one Black and one White, unknown to one another, were selected by the client, who was a White female. The client had a unique relationship with each of the consultants. The White consultant was associated with a gestalt organizational work institute that the client valued, having been involved in some training aspects, while the Black consultant had worked as an employee for the client at an earlier time and had maintained a special colleague/friend relationship with the client over time. The client represented the initial link between the consultants, and a major aspect of the consultation involved the development of the consultants' relationship in order to design and accomplish the work.

In this small, traditionally southern town, with its large medical center, the invasion of a salt-and-pepper (Black and White) consultation team authorized by a relatively new, northern administrator was viewed with caution and suspicion. As in most hospitals, the hierarchical structure presented White males at the pinnacle, in roles of administrators and physicians, and Blacks or other minorities at the bottom, in roles of custodial workers and unlicensed assistants. Within the nursing department, which was the primary site for the consultation, the hierarchy was reflective of this same demographical sketch, except that White females were at the top, in administrative and licensed caregiver roles, and Blacks occupied the bottom rung of the ladder.

Historical aspects of the relationship between the Black consultant and the client could be described as caring and supportive, with the Black consultant representing the primary caregiver. At the beginning of the early work relationship between the Black consultant and the client, the Black consultant had been in a consultative role with the client and was then hired as an employee of the client. While the contractual and payment mechanisms changed, the Black consultant remained in a partial consultative role even in the new position of employee as the supportive relationship established prior to official employment overflowed into the daily work environment. In addition to other standard work expectations, the Black consultant was the client's primary confidante, guide in terms of sensitive communication issues, and sounding board concerning career goals of the employer/client. Dumas (1985) characterizes this relationship as reflective of the "mammy" phenomenon, in which Black females are conceptually and operationally managed by Whites as the all-giving plantation mother. By splitting off the Black female's sexuality, competence, and other attributes of a whole person, the Black female is reduced to less of a threat to Whites. Interestingly, this dynamic was played out by both the client and the White member of the consultant team in each one's relationship with the Black consultant. Obviously the Black consultant had a valence for this role and the rewards as well as the losses associated with it.

The rewards for the mammy role ranged from feelings of superiority and proving that Black women are to be valued to the personal glow from being on the vulnerable inside of those in positions of power and authority. The losses involved the absorption of one's time and energy, since the work expectations of the job are not eliminated while the other supportive, caretaking activities are delivered. This loss of time and energy may result in poor performance in meeting the work expectations, with the concurrent loss of feelings of competency. The loss even extends to the absence of a support/caring network or individual for the Black woman who is caught in the "mammy" complex. The lack of reciprocity at some point becomes chilling and unacceptable.

With this context as background, the chapter will focus on a consultation that was a rescue mission with the salt-and-pepper consultation team as the savior for the client. The overt agreement involved a consultation that could be primarily described as a planning process to determine the need and direction for a more intensive, long-term consultation. Three major periods will be described: Preassessment, Assessment, and Postassessment.

## PREASSESSMENT

### The Request

The consultation was initiated at the request of the client who was chief nursing officer (CNO) of an academic medical center. The medical center had a sterling reputation at both the national and international level. Located near a small urban area with a high proportion of African Americans, the medical center was extremely sensitive to issues of differences and race. This sensitivity had been heightened by a series of media attacks concerning the employment policies and alleged racial injustice of the institution. The CNO had been hired approximately 2 years earlier after competing with an interim CNO internal candidate for the position. The interim CNO was a White female southerner, a native of the area, while the present CNO was from the North. Part of the dynamics played out included the relationship differential between northern Whites and Blacks as compared with southern Whites and Blacks.

The competing interpretations were northern Whites had better, more respectful relationships with Blacks than southern Whites; in contrast, southern Whites had more intimate and caring relationships with Blacks than northern Whites. The power of these different interpretations comprised part of the tapestry of the dilemma for the CNO, a northern White who viewed herself as consciously respectful of blacks and culturally sensitive to issues of differences. This view was reinforced by the White, northern member of the consultation team. The opposing interpretation was carried by the majority of the nursing administrative team and the Black staff members. The Black consultant, a southern member of the consultation team, held aspects of both interpretations, having been raised in the South, but living in the North for most of her professional career.

The medical center had endured several years of existing without a consistent leader. During the 2 years of the CNO's employment, three different chief executive officers (CEOs) had been in charge. One had been pushed out, one died, and the third had just arrived. During this time period, leaders were "executed" in a variety of ways. The possibility of our client's execution queasily rested in the group mind of the institution and the consultation team, while the question of the consultants' possible role in the condemnation and removal process of the CNO was not discussed. Perhaps the verbalization or discussion of this potential outcome was feared as a pronouncement of imminent failure for the CNO, the consultants, and the resolution of the issues within the organization. The "dance around the edges" of critical issues would prove to be thematic to the consultation.

Prior to the consultants' arrival, several major assessment strategies and interventions were undertaken. Two surveys were completed by the nursing department staff, which was primarily composed of White females in administrative and professional positions and Black females with some males in paraprofessional or technical positions. Both surveys indicated that many of them, regardless of race, felt alienated. Results from a management effectiveness analysis instrument administered to all management staff indicated a need for empowerment skills: specifically group dynamics, effective delegation, motivation of staff, and management of conflict and change. In addition, an ongoing intervention, which had no definitive time boundary, involved the consultative work of a psychologist, a northern White female similar to the White member of the salt-and-pepper consultant team, whose purpose was to ease the transition of the new CNO with her administrative team.

With the addition of our two member consultation team, more than five consultants had been brought into the nursing department over a 2-year period. The continuous flow of those who were paid to assist was viewed as a testament to the inadequacy rather than the competency of the CNO. Prior to the arrival of the new CNO, consultants were rarely used in the nursing department.

The preassessment phase brimmed with diagnostic indicators of major issues that resounded throughout the consultation. From the assembling of the consultation team to the actual charge and expected outcomes, the issues of power, authority, safety, change, race, and even the pace of change were available to the discerning eye. The consultation process was distinguished by the control exercised by the CNO, as two consultants without any previous work experience with each other were brought together to blend their skills and knowledge in a rapid-fire, changing southern hospital environment. On the other hand, the opportunity to act as if race was a coincidental factor, not primary or central to the consultants' work, was seductive. The Black consultant's unspoken wish was to relegate race to the category labeled coincidental, whether referring to the consultation or to the personal life of the consultant. The color/race issue was highlighted simply by the presence of the two consultants. We could all pretend this was the extent of its significance. This as-if fantasy was parallel to the whole southern medical center system's denial of race being a significant issue in the interplay of roles in the care delivery institution.

## THE ASSESSMENT

### The Stated Charge

The charge was identified by the CNO as follows: (1) Assess the nursing department's need for team development efforts that could enhance the department's ability to work collaboratively as well as to manage differences in ways that facilitated quality patient care, and (2) offer recommendations to the nursing directors to assist them in developing an implementation plan based upon the assessment. These recommendations were to be offered during a planning day for the nursing directors, the CNO's administrative team. The consultants were asked to serve as facilitators for the planning day.

### The Unstated Charge

The salt-and-pepper team was brought in to verify the CNO's commitment to diversity as well as her proficiency as an administrator, thereby reinforcing the institution's wisdom of hiring her. For the CNO, this confirmation process would have the advantage of identifying potential disloyal, competitive challengers to the throne. This could be achieved through the detailed interviewing of the CNO's leadership team by the consultants, both clinicians. Simultaneously the interview process could identify character witnesses or supporters of the CNO. The unstated charge was to rescue the CNO from an upcoming execution, resulting in her elimination from the system. While the unstated charge was not clearly, consciously acknowledged by either consultant from the beginning, the number of preceding consultants to the nursing department and the composition of the consultation team indicated the CNO's sense of desperation to survive the system.

## Methodology

An intensive interview strategy was used as the primary methodology. The executive administrative team along with the head nurses were divided into two groups, and each individual was interviewed separately by one of the consultants. Interview time ranged from 1 to 2 hours. An extensive interview time, approximately 4 hours, was spent with the CNO by both consultants. After the initial interviews with the CNO as well as other members of the administrative team, time with leadership became rare. Instead, there was the pretense that the issues related to the charge resided in the staff, with the administrative team viewed by the CNO as a group of cohesive yet puzzled observers.

The consultants were aware of underlying, throbbing dynamics in the CNO and her administrative team, but were conveniently distracted by their own intense involvement in developing a working relationship around issues of competition, competency, race, and sexuality. The White member of the consultation team revealed that her lover was a Black male with whom she had a tumultuous relationship. The Black consultant, on a personal level, immediately reacted to the scarcity issue of "good" Black men for "good" Black women. As the plea for help and lovelorn consultation erupted from the White team member, the Black team member angrily stewed over the inappropriate use of that scarce resource, the Black man. But always seduced by a cry for help, the Black consultant became a listening ear for the Black-White relationship struggles. This capitulation by the Black consultant to the comfortable role of "mammy" proved an adequate distraction to maintain the illusion that the issues requiring discovery resided at the staff level and did not necessarily include the CNO or her administrative team. By exonerating the administrative level from scrutiny, the potential of an execution of the CNO was avoided by both consultants.

## Results of the Assessment

The need for vision, direction, and stability in the aftermath of multiple CEOs and the major CNO changes were the most frequently mentioned challenges. Repeatedly, the respondents made the following types of comments: "short time frames and constant crisis"; "heal the past; the fog of the past clouds the present." These were initial topical statements that only suggested the potential depth of despair, anger, and rage experienced by the majority of the staff. While managing conflict and race relations were embedded in the interstitial spaces of the staff's responses, there was no direct acknowledgment of race as an issue.

## Issues

It was during the in-depth interview sessions that the curtain was politely drawn back and the intensity of the pain existing in the organization became apparent. To the unconscious relief of both consultants, the pervasiveness of the organizational issues was so compelling that on-site time was cleanly absorbed, leaving minimal time for interacting face to face with one another or with the CNO. The salt-and-pepper team had moved from entering the system as singles, to a threesome including the CNO, to a pair with the CNO excluded, and finally back to singles. As a result, issues of competition between one another or with the CNO were not discussed. The process of eliciting traumatic experiences from the staff served to dismiss the need to struggle with further development of the consultation team or with the dynamics of the leadership. The salt-and-pepper team's attention was externally turned to the collection of issues from others, as if the focus on others relieved the team of its responsibility for being part of the lived experience, which included digesting the angry reality of the absence of the CNO's presence and authority in the daily work environment, as expressed by the staff and administrative team. The following issues represent themes or patterns gathered from the assessment process.

1. Pervasive distrust—This distrust permeated all levels: head nurses with directors, staff with administrators, director to director and director with the CNO. It clearly extended beyond nursing, but the consultation contract did not include exploring and crossing the boundary into the larger system.

2. Miscommunication and missed communication—With an institutional history of using secrecy as a covenant to those belonging to a certain segmented network within the organization, missed communication was the price of being outside a viable network. Networks were constructed according to geographic origin (North—outsiders vs. South—insiders), race (Black vs. White), administrative level (manager vs. staff) and tenure in the institution (less than 10 years vs. greater than 10 years). With information being used as a power tool to reward and punish, the following strategies were frequently used:

- A reliance on memos to convey information that would be more clearly received if conveyed by a combination of written and verbal communication.
- No advance warnings of changes to individuals who will be responsible for or affected by the follow-up.
- The use of the grapevine as a primary source of information.
- Networks used to sequester information.
- Late notification of meetings and activities.

3. Uncertainty and inconsistency—While there were positive changes occurring in the nursing department, the changes challenged and diminished the institution's traditional identity. Amazingly, the old traditional institutional identity was often described by the majority of networks as a plantation. Whether the plantation was viewed as gracious, good, polite, and a sheltering umbrella for all or as a master-and-slave existence seemed to vary with the racial identity of the network. The Black network viewed the institution as the harsh, hypocritical side of the plantation and the White networks seemed amenable to the gentler aspect of a plantation for the institution. All the networks agreed that the ground was shifting, and the players were retreating to their

informal power bases within their networks to hold on to some level of certainty. The change process seemed to have no ending or punctuation; one change flowed into the next. Management seemed as victimized by the constant flow of change as the employees. Even a totalitarian system of masters and slaves was more acceptable and more easily managed than the evolving unknown.

4. Issues of favoritism and loyalty—Consistent with the psychology of plantations and dynasties, those within the inner circle of power tend to be more valued, rewarded, and assured of their positions within the organization. The admission price to the inner circle seemed to be loyalty, but not necessarily to the CNO, but rather loyalty to the traditional institution, to the plantation concept. Through the issue of loyalty, racial divisions were breached as the insiders, both Black and White of all socioeconomic ranks, came together against the outsiders, which included the northern CNO.

On the other hand, favoritism was the total province of the CNO. Several of her favorites included loyalists to the old regime who were indirectly competing with her for power. One particularly challenging contender was a protégé of the previous interim nursing administrator who had pursued and lost the CNO position. This protégé served as an unrelenting critic of the changes and an ever-present supportive voice of the traditional nursing leadership against the CNO. The CNO's response was to relentlessly pursue and perceive the protégé as an active, loyal administrative team member. This response was consistent with the theme of denial that the administrative team was conducting a low-key revolt to protect the South from the carpetbaggers, the CNO and consultants, from the North.

5. Managing differences: Southern style—With the institution being located and embedded in the deep South and its traditions, and with the preponderance of the nursing department being female, the management of conflict was rarely an overt process. Conflict was nurtured, hidden and, if necessary, addressed in sequestered networks through indirect modes, such as requesting intervention by others—male administrators, physicians, or someone higher in the hierarchy. Differences tended to remain underground beneath an aura of politeness and social

charm. Issues, with underlying primitive feelings of envy and rage, did not surface until explosive. Staff reported that retribution for acknowledging differences and conflict usually followed a rule of thumb, "Wound, not kill." The idea of retaliating nicely remained paramount as a Southern tradition. However, wounding was seen as effectively emphasizing one's power, a soft and consistent reminder that the wound could have been a "kill" thrust.

In order for differences not to be noticed, there was even identified an attitude of anti-achievement among a certain network of employees. Two individuals who were successfully moving with the changes were presented with awards by management, only to be criticized and heckled by their peers for their accomplishments. In this traditional Southern organization, it was safer to be one of the crowd than to stand out by being different.

6. The visible difference of color—Steeped in the tradition of segregation with separate cafeterias, bathrooms, water fountains and wards, the institution had not outgrown its past. Within certain hospital units, on certain shifts, the staff remained all Black or all White. Many 30-year staff members had personal memories of the Jim Crow days and the two-tiered system that reflected the passion of a Civil War that in the hearts of many was never settled. One employee's statement captured the spirit of this dialectical situation. "At night, the hospital turns 'Black' as all the Whites go home and the staff, including housekeeping, are mostly Black."

The power of the Black networks was community-based. The institution was adjacent to a predominantly African American urban area. The Black staff of the institution developed an underground railroad from the community into the hospital and frequently used the community media as its most powerful weapon. Rather than an employee dealing directly with the institution about an issue, the community newspaper highlighted the tyrannical role of the hospital in problematic employee situations with racial overtones.

7. Rhetoric versus reality—While the academic center had the reputation for being one of the most prestigious health care institutions in the country, the delivery of basic nursing care suffered. Changes were being orchestrated for complex new systems

of care delivery, yet the basic implementation of care, in need of improvement, was being ignored. Spots of excellence existed, but quality of nursing care throughout the department was inconsistent. Changes to improve care were frequently discounted, as employees viewed the leadership as transient, with the basic motto: "This too shall pass." The passive-aggressive nature of this response is reflective of the tight, protective wall of indifference and yet rage that seemed to protect and simultaneously damage the staff.

In this stream of urgent activity, consultants were used to help in the change process to relieve some of the workload through reorganization of the work. However, the input from staff needed to move the consultants up to speed was seen as increasing the time demands. Staff, both Black and White, described a parade of consultants as "foreigners." For every consultant, there was a retreat, with the related assigned tasks for new outcomes not necessarily related to those outcomes identified earlier in the month.

Change was also viewed as a predator of autonomy. While staff identified the need to upgrade the quality of care and improve efficiency by having uniform expectations and applications across functions and units, they expressed a greater need to retain some autonomous right to meet the unique needs of their patient population and staff. Following the extended gap in leadership which had been described as the "decade of emptiness," the new CNO's primary goal was to centralize and standardize goals and unit outcomes. This pendulum swing unintentionally increased resistance to any change, even if the change was perceived as good, since change was equated with the loss of control.

## POSTASSESSMENT

With the completion of the assessment, the salt-and-pepper team hit the wall of reality. This reality demanded an unvarnished description of the despair of the work environment and the workers, the smell of revolt and revolution that pervaded the hallways and office spaces, and the sense that the CNO's time was running out and that someone from her administrative team had been delegated the job of executioner. While this undiluted information was shared with the CNO, her own strong denial system ruled out and transformed the reality to an acceptable form. This fantasy suggested that with additional consultation the organizational crisis could be reframed with a secure positional outcome for the CNO. For the CNO, it seemed as if admitting the reality of the system's momentum in reconstructing the leadership structure without her invalidated her personal worth and competency as an administrator. From the consultants' perspective, while the CNO's attributes played a part, it was a larger systems issue than the personal characteristics and competencies of the CNO. The system had used her leadership to refocus on the traditional values of southern life and culture, to build bridges across racial and positional networks in order to survive. These bridges were not based on a greater understanding or respect for one another but on the collective wish to survive in their valued traditional ways. However, the CNO could not be persuaded to consider the reality of the assessment and the prognosis for her survival as CNO.

The Black consultant's personal sadness was extensive. The salt-and-pepper consultant team concluded that the final step in the consultative process would be to present to the CNO and administrative team the assessment and a possible plan of action for consideration. Both consultants expressed a feeling of finality, indicating that the consultation would not continue, and the planning day with the administrative team was more form than substance. Based on the assessment, several options were presented as if there was a commitment from the administrative team to continue to work with the CNO. The proposed plan was as follows:

1. Develop, articulate, and publish a vision statement that includes a united nursing leadership and nursing department.
2. Involve a cross-section of the nursing department, using variables of race, position, geographical origin, and age to constitute work groups in validating the vision and its implementation.

3. Slow the continuous flow of change. Punctuate the change process with opportunities for rest and re-equilibration.
4. Integrate the changes throughout the nursing system by creating safe forums for discussion and both internal and external opportunities for leadership and staff growth.

## FINAL COMMENTS

Approximately 6 months after the consultation, the CNO was fired and the majority of her administrative team was outsourced or reintegrated into a newly structured nursing organizational system. The primary antagonist from the administrative team identified during the assessment became the new CNO. A leftover question that challenges the consultative mind is how much did the salt-and-pepper consultation team contribute to the dissolution of the existing system and to the demise of the CNO? As a member of the team who carried, perhaps for the organization, the affection for the CNO, the need to prophetically warn the CNO of forthcoming doom was wrenching, and the need to let go of the Black consultant's salvific rescue fantasies was equally traumatic. It was as if one had entered a situation with a friend hanging from a cliff, and at the end of the work, the friend remains dangling from the same cliff yet refusing to admit the impending doom inherent in the scenario. For the Black consultant to walk away from a cherished colleague in a serious state of denial was incredibly difficult. The nurturer, mammy, nurse and caregiver roles were colored with the Black consultant's perceived inadequacy in fulfilling them through a dramatic rescue consultation intervention. Likewise, the "caring" roles with the White partner were fragmented and elusive, as the consultation was concluded without any resolution to her personal relationship with her Black lover or to the development of the work relationship between the Black and White consultants. The lack of a satisfactory ending to the work was reminiscent of the lack of a satisfactory conclusion to life events. At times consultant roles are lauded as well protected from feelings of responsibility, since consultants provide options and clients ultimately make the choices. The responsibility for the outcome of the system's work resides squarely within the domain of the client. Yet the partnership and engagement of human commitment and feelings lead to shared feelings of responsibility. This consultation, due to the personal and historical relationship established between consultants and client, was ripe for serving as an exemplar for highlighting the fragile boundary between caring and consulting. When caring contaminates and diminishes the effectiveness of the work role, the boundary is not being adequately managed by the consultants.

There were several major lessons learned from this encounter. First, beware of consultations that involve friends or individuals with whom the consultants have already developed emotional attachments. Second, in intense emotional consultations, contract with a shadow consultant—a colleague whose organizational competency and objectivity are valued—who will serve as an off-site, advisory agent to the on-site consultant (see Chapter 9, this volume). Third, for consultants of color, be alert to being used in emotionally demanding rescue operations that are hopeless from the initiation of the consultation. The prognosis for this consultation could easily have been determined from the composition of the consultation team. Two female consultants, one Black and one White, unknown to one another, were selected by the client. Prognosis: Poor. While the CNO verbally denied that the system was falling apart and her leadership was in jeopardy, she, perhaps unconsciously, formed a comforting yet contained Black-and-White rescue team that would predictably fail. This initial decision of consultant selection by the client and its endorsement by the consultants set up a predetermined limitation on potential effective outcomes of the work. Therefore, the consultants' work would contribute to a continued state of denial that the leadership and the system valued dearly. As the Black member of this salt-and-pepper consultation team, the author was particularly sensitive to the racial issues that pervaded the setting of the consultation, but was insensitive to the barriers to working the racial issues that were created by the consultation team's composition. This "color"-blind spot was magnified by the personal and longtime professional relationship of the Black consultant with the client. Part of the seductiveness of these types of consultations might include issues of race, anger, and sexuality. Finally, know your teammate before being paired with a fellow consultant in an explosive, draining consultation effort. The initial development of a foundation for working with a fellow consultant should not be a distraction for remaining on task with an ongoing consultation.

This chapter has been incredibly difficult to write. As the author and the Black consultant, my personal exposure seems immense. My personal pain related to the consultation is fresh and the wounds remain open. The motto "wound, not kill" crossed the permeable boundary between the consultants and the client. Yet with every consultation, there is additional growth and some level of acceptance of the human frailty that even consultants cannot escape. For even consulting roles do not afford protection from feelings and the human experience. While these feelings need to be managed, it is their integration into the consultants' work that produces the highest level of productivity. To the consultants and the wanna-bes reading this book, I wish you continued growth.

## REFERENCE

Dumas, R. G. (1985). Dilemmas of Black females in leadership. In A. M. Colman & M. Geller (Eds.), *The group relations reader 1* (pp. 323–334). Washington, DC: A. K. Rice Institute.

# 5

# Coping with a Divestiture:

## The Psychological and Managerial Dilemmas of Ending Personal, Work, Organizational, and Community Relationships

*Laurence J. Gould, Ph.D.*

Initially, I was approached by the Vice President of Human Resources of a diversified energy company—COALCO[1]—to assist them in coping with the consequences of an ongoing threat of divestiture by the parent company, TG (the Transport Group), a large international conglomerate. For about 2 years, on and off, divestiture was under active consideration, but each time, for a variety of reasons, the decision was put off. Not surprisingly, I found a company under serious stress and in a considerable emotional quandary. This chapter will briefly describe the sequence

---
[1] To ensure confidentiality all names have been changed, as well as some details of the organization and its background.

of events that led to considering an intervention, the rationale and methodology for the intervention, and the outcome.

The theoretical focus of the case is on how the experience of failure and loss, the impact of unacknowledged mourning and grief, and the mobilization of powerful social defenses against persecutory and depressive anxieties resulted in maladaptive, unconscious group and organizational processes. These processes, in turn, made it increasingly difficult for the organization's leadership to constructively develop a clearer idea of how to conceptualize and carry out the required socioemotional, operational, and managerial tasks during a period of great uncertainty about the future.

Specifics aside, the general issues raised by COALCO's divestiture are quite common, as similar organizational upheavals such as mergers and acquisitions, downsizing, and the wholesale elimination of divisions, units, and product lines become increasingly frequent. In this connection I will try to draw out a few lessons for organizational leadership in the 21st century.

## THE COALCO ORGANIZATION: PAST AND PRESENT

COALCO was established in the early 1900s as a partnership, but over the years it became the family business of one of the original founders. The modern history of the company began in 1948, when the son of the founder, Henry Witt, was elected president—a position he retained until 1977, when the family sold the company to TG, which had hoped to develop a strong position in the energy industry at a time when it looked like coal was going to replace expensive and unreliable oil imports. During the period of Mr. Witt's tenure as President, COALCO became one of the largest and most progressive coal companies in the United States, and he became one of the leading citizens of Kentucky and a national leader in the coal industry. It is by no means a minor addendum to this story that Mr. Witt, who lived until the age of 88, bequeathed a gift of TG stock worth over a million dollars to the Office of World Hunger of the Christian International Assembly "to be used for aid and assistance to the hungry and needy, at home and abroad." These funds later provided food and medical services for children and families in need, from rural Kentucky and West Virginia to the Philippines, Mozambique, Zimbabwe, and Malawi.

This history of care, community service, and charitable pursuits was still very much in evidence in COALCO's culture. It is hardly surprising that having to manage the process of divesting the company, thus ending a 90-year legacy of "good works" and the company's illustrious name, as well as what had been a very successful business, was a terribly painful ordeal, fraught with more than the usual anxiety, guilt, shame, and a sense of failure that such endings stimulate under the best of circumstances.

## PLANNING THE INTERVENTION

After a series of preliminary discussions and individual interviews with each member of the senior management team, a decision was made to develop a venue to discuss the impact of the impending divestiture on each individual, and on the company as a whole. In the words of COALCO'S president, reflecting, some time later on this decision

> It resulted from a series of events which began last year and involved my perception that in the course of selling the company there were a significant number of issues relating to our individual relationships, and our relationships to external groups that would become stressed to their limit during the process, and that perhaps some type of organized program to develop our individual and collective thinking about these elements would be useful.

Management hoped that such a program would reduce stress, lead to greater sharing and openness about anxieties regarding the future, enhance the spirit of colleagueship ("we're all in this together"), and clarify the conceptualization of the tasks of senior management during the divestiture process. After deciding to go ahead, and setting a date for a retreat—about 2 months following the initial contact—we also agreed to hold a follow-up meeting, approximately 3 months hence.

Three weeks before the retreat, the parent company, TG, finally decided to actively proceed with the divestiture and put COALCO up for sale. After a brief discussion it was agreed that the purpose of the retreat and associated activities, as originally outlined, had not changed except for the immediacy of the concerns. Therefore, it was felt that a retreat would be a useful format for the senior management to discuss the individual and organizational impact of the divestiture, and to do some collective planning in light of it. Further, the senior management would now have to be actively involved in preparing for, planning, and managing the divestiture process itself. In effect, they were responsible for organizing their own funeral! This, of course, raised extraordinarily complex emotional and managerial issues. The overall purposes of the retreat, therefore, as offered to senior management for its sanction, were conceptualized as follows:

- The primary task of COALCO's senior management, at this point in time, is to facilitate the divestiture process, resulting in a positive outcome.
- In light of the above, each member of COALCO's senior management team, and the team as a whole, needs to begin to work at the task of constructively ending relationships with the company, each other, employees, customers, the parent co., the community, etc.
- The overall task of the retreat, therefore, is to develop the thinking and actions that contribute to the above.

The overarching rationale for stating the purposes of the retreat in this manner was to emphasize that there was a dual set of tasks: that is, a *content* or substantive task, and a *process* task. This description is similar to how Bridger (1987) and his colleagues conceptualize their training events. In the case of COALCO the substantive task was bringing the divestiture to a successful and expeditious conclusion (Task 1), and the process task (Task 2) was to explore the feelings and behavior that the divestiture produced, with the goal of facilitating more constructive endings as well, and to determine how these feelings affected the accomplishment of Task 1. The idea that I put forth, and repeatedly emphasized, was that how relationships are ended is as important, if not more so, than the fact that they end.

## THE INTERVENTION: AN OVERVIEW

A process was designed to provide a framework for addressing the preceding issues in a systematic fashion. It consisted of four interrelated activities: in-depth interviews with each member of the senior management team, the off-site retreat mentioned earlier, a follow-up meeting, and some ongoing consultation that included providing a variety of relevant readings and sending along notes, letters, and news clippings that I thought would be germane to their situation.

### Interviews

Prior to the interviews, each member of the senior management team was asked to prepare, in writing, an assessment of the situation. To gather this information, they filled out a preretreat worksheet that solicited their views about COALCO and its relationships with significant internal (e.g., employees at different levels) and external constituencies or stakeholders (corporations, unions, suppliers, customers, their families, community agencies, schools, etc.). These materials provided the initial starting point for the interviews, in which I invited the managers to collaborate with me in generating relevant data, raising crucial issues and questions, and developing hypotheses about how they—the management group—and COALCO were functioning in light of their situation. I then took the responsibility for organizing this information, and feeding it back to them (in this instance, early in the retreat) as an initial input to the ensuing deliberations. Put in the form of questions, some of the recurring issues and concerns raised are as follows:

1. How can we prepare our good-byes to each other, to our employees, to our customers, and to the other groups with which we have contact?

2. How can we stay mindful of the feelings of the others in the company and in our personal lives who are also affected by the divestiture?
3. How can we care for others when we are not sure they will care for us?
4. What happens if COALCO doesn't sell?
5. What happens if one or more of us leaves early?
6. How can we help COALCO keep its good name and legacy?
7. What do I want from my peers during this process?
8. What are the perceived operational objectives necessary to achieve a successful divestiture?
9. How do we manage our morale and productivity and how do we as a group share responsibility for this?
10. What of value can we take with us as managers when we leave?
11. How do we manage internal conflict and competing agendas among ourselves during the divestiture process?
12. How do we manage ongoing but dynamic business concerns that may appear to conflict with the divestiture (i.e., short-term and long-term productivity improvement versus the divestiture goal)?
13. How do we manage to instill a concept of the "future" for ourselves, key managers, and for various work groups with diverse skills and opportunities?
14. How do we develop a framework for combating organizational paralysis centering around the underlying psychology of the divestiture?

## The Retreat

The retreat was held over a period of 4 days at an off-site facility. The major activities centered around sharing experiences about the divestiture and ending relationships, working on strategies to constructively facilitate the divestiture, and developing action plans to move the process along—the divestiture itself, as well as an exploration of their relationships, and an ongoing collective self-assessment of the state of their team. A series of evening activities were also planned to sensitize them to some of the issues they and the staff were struggling with, and to stimulate a deeper and more personal level of engagement. These included watching and discussing two films—*Matewan* (Sayles, 1987), and a documentary of a real-time participative discussion with a group of recently unemployed Canadian workers, titled *Who's in Charge?* (King & Lawrence, 1982)—and a presentation by and discussion with the co-director of the Appalachian Rural Housing Authority.

As noted, the feedback of the interview data provided the initial basis for discussion at the retreat. Over the course of the following three days these data were elaborated and developed in the context of three other major activities:

1. Developing hypotheses about the nature of the management team, and how it was functioning.
2. Giving individual feedback to members of the team about how they have been experienced over the years.
3. Providing an opportunity for all members of the team to develop and present to the others their personal action plans for addressing the issues of concern in connection with their own staff and functional areas of responsibility, the tasks required of the organization as a whole, and their families, friends, and neighbors.

Before discussing the dynamics of the management team and how it functioned, I would note with regard to giving individual feedback that effective management development—specifically, individual development—and organizational development go hand in hand. Although all human resource development efforts must be primarily aimed at supporting the organization's strategic and tactical objectives, it is very useful to provide, if possible, simultaneous opportunities for the individual's professional growth and development. Of course, in this situation, with the company's days numbered, life after COALCO and what of value one could leave with were very much on people's minds. Therefore, as the explicit rationale for having them share their views of each other, I developed three hypotheses, which I presented and discussed with them as the context for this activity:

1. Candid and constructive feedback is a gift.
2. Candid and constructive feedback will be of use in both the current situation and in the future, whatever the outcome.
3. Offering each other candid feedback is an important part of the process of constructively ending relationships.

## THE DYNAMICS OF THE MANAGEMENT TEAM AND HOW IT FUNCTIONED

Utilizing a methodology called "sculpting," originally developed in family therapy work, the participants were asked to array themselves in a group as they experienced themselves in relation to others. Specifically, they were given the following instruction:

> The objective of this task is to enable members of the COALCO's senior management to locate themselves on the team.
> Please take a chair and sit in a position which reflects where you see yourself in relation to the other members of the team.

I then asked the participants to volunteer their understanding about the configuration they had created. After several participants spoke, I asked whether anyone had a different idea about how the relationships might be more accurately depicted, and asked them to reconfigure the group accordingly and to describe why they had done so in this manner. Following a number of iterations, the group was then divided (with their input) along the subgroup boundaries that they had created, and they were asked to develop hypotheses about how the team functioned. Each subgroup then reported its thinking as the basis for a subsequent discussion. The extent to which the collective self-appraisals were consistent across subgroups was quite striking, and certainly consistent with my own developing hypotheses about them. I would emphasize here that having a group appraise itself, at least initially, rather than my interpreting their behavior from the outset, considerably reduces the resistance to subsequent painful insights, and facilitates the responsibility for and ownership of what they have created together. It also considerably increases the likelihood of the group's developing constructive strategies for addressing its difficulties, as I hope the following material illustrates.

The major dilemma, as they articulated it, was the tendency in the group toward withdrawal and consequent isolation. The first of the three subgroups said, "This exercise erroneously assumes that the eight of us here comprise one team"; the second said, "No team as a whole exists"; and the third said, "We are unwilling to sort issues for teamwork." Perhaps it goes without saying that the dynamic of disengagement made it extremely difficult for the group to mount the sort of collaboration required to address either the socioemotional or the operational and management tasks at hand, since it led to behaving *as if* business would simply continue as usual. When the group arrived at this point of recognition during the discussion, I was then able to say that I felt that they were behaving "like orphans-to-be" who could only wait helplessly for their parents to die, but could not tolerate or share with each other that painful idea. To the extent that they could link up with this interpretation, I was also then able to point out that, in fact, they could potentially have a considerable impact on who might buy COALCO, since they were the ones essentially showing the company, and they, not the parent company, had relationships with and knew all of the potential buyers. This idea galvanized the group into action, and during the lunch break they organized themselves to systematically draw up a list of all potential buyers, did an initial evaluation of the pros and cons of being acquired by each, and brainstormed what they could do to tailor their sales pitches accordingly. The lively discussion generated by this activity considerably lightened the mood and informed their subsequent work.

## CONCEPTUALIZING THE GROUP DYNAMICS OF COALCO'S SENIOR TEAM

From the framework of my conceptual perspective—mainly Bion's *basic assumption* theory—I offer several hypotheses about COALCO's dynamics drawn from the episode briefly described, as well as from data that emerged from other observations and discussions. I also comment briefly on the nature of the coal industry and its contribution to these dynamics, and how the particularities of the completion bonus package exacerbated the difficulties experienced by COALCO'S senior management. Before doing so, however, let me briefly outline Bion's theory.

## Bion's Basic Assumption Theory

Since Bion's (1961) *basic assumption* theory is well known (e.g., Rioch, 1970; Turquet, 1974; Lawrence, Bain, & Gould, 1996) I will not review it in detail. Suffice it to say, therefore, that the *basic assumptions—ba Fight/Flight, ba Dependency,* and *ba Pairing—* may be viewed as central modes of group behavior that coalesce around different patterns of drives, affects, mental contents, object relations, and defenses. That is, when Bion speaks of a basic assumption group he is not talking about its members, but rather the mental state of the group characterized by a particular constellation of affects and processes driven by a need for emotional security. These basic assumption modes are contrasted by Bion with another mode—the *work,* or *W,* group—the aim of which is task performance. A useful, and, by now, conventional way of understanding basic assumption states is to regard their operation from an *as if* perspective. For example, when a group is a *ba F,* one would say "it is acting *as if* its survival depends on *fighting* (or *fleeing*)." From this vantage point, survival drives all of the basic assumptions, and can be viewed as the meta basic assumption; "The group is met to survive." Finally, it is Bion's (1961) view that a preoccupation with a particular basic assumption "leads a group to ignore other activities, or, if it cannot do this, to suppress them or run away from them" (p. 64). Prominent among these are activities and processes related to the *reality principle,* a full appreciation of the external world, and the use of "scientific procedures"—in a word, activities prominent in a *W* group.

## The Group Dynamics of COALCO's Senior Team

From the vantage point of Bion's theory, the COALCO senior team, as it initially displayed itself, was largely in *ba Fight/Flight (ba F)*, with collusive denial as the major social defense (Menzies, 1967). However, it also became clear that *ba F* had increasingly displaced *ba Dependency (ba D)* as the dominant mode, and oscillated with it in increasingly longer cycles. My understanding of this particular dynamic outcome was that it was overdetermined by the activation of two processes. First and foremost was a sense in the group of the president's failed leadership, and the feeling that one could no longer depend on him. After all, in their eyes he was not able to prevent the divestiture, and in that sense, in his role, he was the obvious candidate for containing the disowned and disavowed experience of failure projected into him by the rest of the management team. Second, and more peripheral, was his *pairing* with the vice president of Human Resources—the only woman in the group—which became a source of irritation, as others felt increasingly excluded and out of contact with him. Taken together, these feelings led, I believe, to an erosion of any semblance of sophisticated *ba Dependency (ba D)* and, as a result, the president's leadership could no longer contain the anxieties mobilized by the impending divestiture. This process then resulted in a regressive shift to *ba F*, with the denial and disengagement already noted, as the dominant defensive modes. The group's self-appraisal of these difficulties, deepened by further interpretative work, facilitated a sufficient *working through* of some of the painful feelings, and allowed more adaptive behaviors to be mobilized and carried forward (i.e., the assessment of potential buyers).

## The Nature of the Coal Industry and Its Psychic Manifestations

Several other processes, I believe, are significant contributors—both psychically and organizationally—to an in-depth understanding of COALCO's situation. First, in the realm of mental processes, the nature of the coal industry itself provides, under any circumstances, an overabundance of psychic fuel (choice of words intended) for a variety of persecutory and depressive anxieties as conceptualized by M. Klein (1940, 1946). From having to manage constant accusations of being environmental despoilers, to being accused—at least historically—of killing hapless and helpless miners via either black lung or cave-ins and other mining disasters, it is understandable that the level of unconscious guilt and self-hatred in the industry is quite high, and that the threshold for depressive reactions is, therefore, quite low.[2] The corollary, of course, is that powerful and ubiquitous social defenses are constantly being mobilized to keep the psyche clean (again, the choice of words is intended), since, for example, it is not surprising for black lung and black psyche, in fantasy, to become an easily mobilized unconscious equation.

However, powerful and primitive defenses—like denial, splitting, and schizoid detachment—have extraordinarily high adaptive price tags, especially in the long run, since they create their own serious organizational mischief. With regard to COALCO's situation, the unconscious guilt that they let their miners and other staff down by not being able to prevent the divestiture, possibly coupled with "historical guilt," contributed to senior management's passivity and disengagement. The guilt about failing to prevent the divestiture did, in fact, subsequently emerge quite clearly.

## The Completion Bonus Dilemma

Organizationally, another factor—namely, the structure of compensation—also exacerbated an already difficult situation. Specifically, TG offered a generous completion bonus to those senior employees who stayed on until the company was sold (this was later modified to termination of services). While not exactly a

---

[2] One might speculate that Mr. Witt's history of supporting "good works" and his generous bequest represented, in part, unconscious guilt-driven reparations.

"golden parachute," the bonus did indeed function like a pair of "silver handcuffs," and made it quite difficult for senior managers to get on with their lives and careers, especially since the divestiture was taking place in a depressed economy, with few available jobs. As the sale process (and the economic depression) dragged on, people felt pressured to remain, even as they experienced increasing depression, helplessness, and despair.

## A NOTE ON THE DEATH OF AN ORGANIZATION AND SOME LEADERSHIP LESSONS FOR THE 21ST CENTURY

Some aspects of my own experience as I worked with COALCO are relevant here because of the implications that I draw from them with regard to leadership requirements in the coming century. Most immediately, I felt extremely depressed and ill at ease, and found myself thinking about and dreaming about COALCO a great deal, even when I was far away from it. As I reflected on this experience, I discovered that it is much easier—no matter how painful—to get one's mind and heart around the death of an individual than the death of an organization. How do we experience such a loss—both those who are part of it and the external stakeholders and the community that transact with it and contain it? Further, what does the community, and by extension the larger society, lose when a company like COALCO dies? I felt strongly, based on the retreat and many other conversations, that *the community itself lost a small measure of value and coherence that COALCO had contributed as one nucleus of social concern and moral authority* (recall COALCO's history). The loss then was not just a matter of painful practical realities, like unemployment, or the erosion of the local tax base—these are, of course, serious—but equally involved psychic dimensions, eroding the hope of those directly involved on the one hand, and weakening the community and social fabric on the other.

Just as interdependence is necessary for the survival of the individual, no organization is "an island unto itself"; the loss of an organization is a loss to the whole system in which it is embedded. We are still quite far, I believe, from having a way of working out the "social calculus" of such loses (see Taber, Walsh, and Cooke, 1979, for an excellent early consideration of such issues). Perhaps the task of better understanding and valuing organizational/community relatedness represents a distinct challenge to society and the leadership of its organizations in the 21st century. Taking up this task is a direct counterpoint to the prevailing parochial perspectives regarding an organization's competitive position and prospects for survival. However, accomplishing it requires that someone with sufficient authority and the wit and wisdom to do so *keeps this task in mind.* Doing so, I believe, will be central to the concerns of enlightened leadership in the years to come. It will require as much appreciation of social consequences as it does of either financial advantage or professional gratification (in the case of not-for-profits), and of process as it does of substance, as well as the ability to contain considerable anxiety, and the wherewithal and courage to lead and work in the socioemotional realm, when required. I would like to suggest here that COALCO's leadership prefigures this important development. How many contemporary organizations would seek consultation to assist them in working at the dilemmas of constructively and responsibly ending relationships?

## THE ENDING

Within a year following the retreat, the divestiture was finally negotiated. No buyer was found, although one of the divisions—the mining machinery company—was bought from TG by its general manager and several outside partners. TG, therefore, decided to "mothball" COALCO, and to transfer the management of its few remaining functions (e.g., fulfilling some sales, union, and environmental reclamation contracts) to one of its subsidiaries. In the words of one of the senior management team, taken from notes of the follow-up staff meeting, subsequent to the retreat,

> We all want a piece of COALCO before it is gone. Who has the right or obligation to protect COALCO, the COALCO legacy? The

divestiture implies failure. Who is responsible? Does it matter who is responsible? We all share the failure. We seldom recognized the successes of COALCO. Did we manage COALCO to this outcome?

Here we see common clinical echoes of unresolved grief and mourning. These, of course, are the enduring questions and concerns that we all face when we come up against death, problematic endings, and failure of any sort—either real or imagined. However, it takes great courage to raise these questions, struggle with them, and fully own the answers, rather than disavowing responsibility for failure, if such it is, and projecting it onto others. In this regard I very much feel that COALCO's leadership and management provided a rare example of such courage, and that it was fitting for a company with a legacy of progressive and humane values to try to work at ending its life constructively and responsibly. For me it was a rare pleasure to work with them. One can only hope that this kind of singular contemporary organizational leadership, though painfully come by and hard won, will become a norm in the 21st century.

## REFERENCES

Bion, W. R. (1961). *Experiences in groups*. New York: Basic Books.

Bridger, H. (1987). Courses and working conferences as transitional learning institutions. In E. Trist & H. Murray (Eds.), *The social engagement of social science* (pp. 221–245). Philadelphia: University of Pennsylvania Press.

King, A. (Producer), & Lawrence, W. G. (Director). (1982). *Who's in Charge?* [Film].

Klein, M. (1940). Mourning and its relation to manic depressive states. *International Journal of Psycho-Analysis, 21*, 125–153.

Klein, M. (1946). Notes on some schizoid mechanisms. *International Journal of Psycho-Analysis, 27*, 99–110.

Lawrence, W. G., Bain, A., & Gould, L. J. (1996). The fifth basic assumption. *Free Associations, 6*, 28–55.

Menzies, I. E. P. (1967). *The functioning of social systems as a defense against anxiety* (Tavistock Pamphlet No. 3). London: Tavistock.

Rioch, M. J. (1970). The work of Wilfred Bion on groups. *Psychiatry, 33*, 56–66.

Sayles, J. (Director/Writer). (1987). *Matewan* [Film].
Taber, T. D., Walsh, J. T., & Cooke, R. A. (1979). Developing a community-based program for reducing the social impact of a plant closing. *Journal of Applied Behavioral Science, 15,* 133–155.
Turquet, P. M. (1974). Leadership: The individual and the group. In G. S. Gibbard, J. J. Hartmann, & R. Mann (Eds.), *Analysis of Groups* (pp. 349–371). San Francisco: Jossey-Bass.

# 6

# The Consultant's Use of the Inner Dialogue in Family Business Consultation

## Joseph Rosenthal, Ed.D., and Donald A. Davidoff, Ph.D.

> *Absolute self-knowledge is, I believe, never a claim of a thoughtful man. The enormous, subjective prejudice that manipulates so broad a field as our memory is only a glimpse of the prejudices and whims that affect our judgment....*
> —*John Cheevers,* The New Yorker

Many of the Fortune 500 companies began as family businesses. With downsizing, restructuring, and changing technology, future jobs will likely be in small entrepreneurial businesses, often family-owned. Therefore, efficient family businesses are critical to the future of our economy.

As Kets de Vries and Miller (1984) point out, transferential patterns manifest themselves in many leadership styles. Leadership styles and decision-making behavior are frequently related to intrapsychic developmental influences that are revealed in transferential situations. They go on to point out that effectiveness of leadership requires an understanding of the less overt manifestations of power and its use, particularly as it applies to interpersonal, group, and organizational dynamics. While acknowledging that recognition and understanding of transferential patterns are not easy to achieve, they note that neglecting such issues can have significant negative effects on a business's ability to adjust efficiently to a dynamic marketplace.

Nowhere is the issue of greater importance than in the family-owned business. It is here that the executive may not only be acting out his or her transference to parents in his or her relationship to the CEO, but the CEO may in reality be that executive's father or mother. The injection of this reality into otherwise intrapsychic processes can prevent the company's leadership from recognizing such noxious influences on decision-making processes. As Levinson (personal communication, September, 1994) suggests, good leaders make good decisions. In family-owned businesses the combination of intrapsychic processes and reality can render the making of "good decisions" almost impossible. The role of the consultant then can be critical in helping the leadership of the family-owned business understand what belongs to the business and what belongs to the family. Negotiating such an understanding can be hazardous for the consultant because of the intensity of the externalized intrapsychic processes.

The task of providing consultation to family-owned businesses presents a series of unique challenges to the consultant. On the one hand, the family business is an external representation of the internal life of the founder. It is thus more susceptible to reacting to stressors in an idiosyncratic manner than a business that is organized purely to perform a particular task. On the other hand, since early family life serves as the individual's template for future participation in all aspects of organizational life, consultation to the family business can evoke powerful resonances in the unwary consultant. McCollum (1990) recognized this unique and reciprocal relationship in her discussion of clinical research on the family business. She defined three important factors that can influence outcomes: (1) The consultant cannot work without being influenced by personal and professional bias; (2) the consultant must view personal emotions as relevant data; and (3) the consultant must strive to develop trust in relationships with system members. While McCollum focuses on the research consultant and cogently points out the wealth of data lost by not taking note of these factors, her ideas are no less appropriate to the role of the business consultant.

It is, however, the rare business consultant who will ask, "What can I learn about the system from how I am treated by it?" and "How do I feel when working with the client system?" and "What does it tell me about how others feel?" Making the choice to entertain this inner dialogue can be experienced as threatening to the consultant for it endangers the "myth of objectivity." Yet the consultant is by no means a neutral bystander, but rather is, in Sullivan's (1953) terms, a *participant observer* in relation to the client system.

At the boundary where the client and the consultant meet, they must join or pair when working together. The consultant has to face the challenge of "becoming an insider" while retaining the qualities brought from outside. While this implies that the consultant must be adept at managing boundaries to serve the "containing" function of the system (Bion, 1961a), it also mandates the necessity to engage in this inner dialogue to make sense of his or her experience.

For a consultation to be effective, the consultant can adopt what Shapiro and Carr (1991) refer to as the "interpretative stance." This stance implies that the consultant is open to experiencing his or her own feelings (as well as thoughts) about a particular consultative relationship. By being able to utilize one's own responses as data concerning the client relationship and the presenting issues, the consultant is able to gain insight not otherwise attainable.

The implications of such a stance are twofold. First, it suggests that such affective reactions, when appropriately understood, are valid indicators of what occurs within a particular system. Second, it implies a certain permeability of ego boundaries on the part of the consultant. Entry into a family business exerts a particularly powerful regressive pull on the consultant because of its essential resonances with family life. Yet this regression is very much in the service of the consultation, allowing the collection of data that would have otherwise remained hidden.

There are four mechanisms capable of activating emotional response in a consultative relationship. These include transference, countertransference, projective identification, and what Grinberg (1979) terms "projective counteridentification." This chapter examines the importance of the consultant's inner dialogue in separating and differentiating these processes. For it is only through understanding what belongs to whom, and how it got there, that the consultant can begin to make sense of his or her internal responses and use them effectively to adopt the interpretative stance. The impact of this interpretative stance of the consultant-client relationship in general and on the issue of boundary management in particular will be discussed.

## THE INNER DIALOGUE: THE CONSULTANT'S USE OF THE SELF

By not attending to an inner dialogue the consultant will not be able to sort out intrapsychic and interpersonal processes. The following example is offered.

One of the authors (J. S. R.) who is, himself, from a family business, negotiated a consultation to evaluate organizational and personnel problems that a family business (the Company) was experiencing. Several discrete areas of concern were to be investigated, including the impending move to a new location, reorganization, downsizing, succession, and business plan, and the relationships among the corporation, customers, and competitors.

The consultant proceeded to interview the various senior managers, which included the owner's children. When asked what it was like to work at the Company, the responses were quite striking in their similarity. The Company was described as a family environment where there was a significant amount of respect for individual initiative and motivation. As a result, this sense of respect was focused on the owner, as patriarch and as the driving force of managerial and product innovation. The owner very much introjected this role and felt himself to be, if not omnipotent, then the only person suited to run the Company. As a result, the owner's children reported difficulty in finding their voice when confronting their father on strategic and long-term planning issues.

Managers and family members alike did, however, respond that they "would go the extra mile" for the owner. To the managers, the owner was more than an employer and mentor; he was described very much like a father. Until recently, joining the Company entailed lifelong membership. For most of the managers this was the case, with many having spent their entire professional lives at the Company. Stories were shared about the kindness the owner demonstrated to employees in need, providing both emotional and financial support.

The managers reported that there was another side as well, that of a company which had lost its way and which was experiencing internal conflict. According to the nonfamily managers, there was an emotional price to pay for belonging to the Company. One of them noted that it was difficult to present a dissenting opinion because doing so would be seen as a direct challenge to the owner. In addition, the managers were unanimously concerned with the health of the owner, noting that if anything happened to him, then the Company would be fatherless as well as leaderless.

Many of the managers reported a helpless feeling and lack of clarity about what performance was expected of the team and the employees of the Company. It was also said that business decisions were often based on the quality of interpersonal, informal relationships and *not* on the formal role relationships.

Thus, the diagnostic process clearly labeled the potential pitfalls for this consultation. Yet in spite of these red flags, the consultant found himself beginning to feel lost in this familiar place and to react on the basis of past relationships and experiences.

To begin with, a consultant is often seen by clients as being akin to a messiah who will lead the company out of chaos and into corporate health. Adding to this basic assumption in *this* consultation was the consultant's own background in his family business, where he had not only been viewed as successor and savior but ultimately had never fulfilled those roles because he had opted out of the business for a career in organizational consultation. This predisposition on the part of the consultant served as a two-edged sword. On one hand, it most certainly enhanced this consultant's ability to empathically and practically understand both the owners' and the Company's needs. On the other hand, it also set the stage for his becoming the target of the owner's rage, which would invariably appear as his omnipotence was questioned.

This consultation was initiated on a positive note, with the owner lauding the consultant for his "deep understanding" of the complexities of family business in general, and the vagaries of the owner's business, in particular. The consultant, in turn, responded to this praise much as a son would respond to a father and expressed the idea that his company "must be a really interesting place to work."

Because of these pleasant, affectively laden interchanges the consultant continued his work on the task he was authorized to execute. As the interviews continued, a pattern became apparent. Briefly, the Company had been losing marketshare, and the employees had become increasingly uncertain about the future. It became clear that the core issue lay in succession of the owner and lack of clarity of the roles of senior managers and family members in the business. The owner had made no plans for his succession and preferred to operate the company on a day-to-day basis with relatively little practical long-term planning.

Once the diagnostic phase was complete, the consultant met with the owner to discuss the findings. As soon as the issues of family involvement and future succession arose, the emotional climate shifted. The owner began to exhibit significant anxiety. He began to become argumentative and emotional at meetings with the consultant. He also began to come late to meetings, walk out of meetings, or cancel meetings precipitously. The owner then rejected the initial findings out of hand. He also rejected the consultant by noting that the consultant was "just too young and inexperienced to really understand the complexities of *his* business."

As the discussion continued, the consultant found himself losing his own voice, feeling chastised and infantilized just as the owner's children had felt whenever they presented an unpopular viewpoint. The consultant, in effect, found himself experiencing a reality not his own, but rather that of the owner's children. Furthermore, the consultant reacted to this shift by resonating with his own past experiences in his family's business. He became equally anxious and increasingly invested in making a case for the role of the children in the Company.

The initial phase of the consultation ended with the owner asserting control over the consultant by insisting that the consultant return to his firm and bring back "one of the senior consultants who is better equipped to understand the business." The consultant felt depressed, angry, and useless. His first impulse, before examining the origins of these feelings, was to do as the owner asked.

At this point, what does the inner dialogue tell the consultant about what has occurred? Initially, the earliest interaction was mediated by transference and countertransference. The patriarch related to the younger consultant as his favorite son and the consultant, in turn, related as if the patriarch were his father. In any novel relationship, it is relatively easy to relate to the principal in familiar ways, the most familiar being modeled on family relationships.

Transference, according to Racker (1968), is the expression of the internal object relations. In other words, in the case of the consultant-client relationship it is the "displacement [by the client] of patterns of feelings, thoughts, and behavior, primarily experiences in childhood, onto [the consultant] involved in a current relationship. Because the process is largely unconscious, the [client] does not perceive the various sources of transference attitudes, fantasies, and feelings" (Moore & Fine, 1990, p. 196). The two factors that have an impact on the relationship between the consultant and client are the real qualities of the interface and of the consultant. These current factors are experiences through the filter of the past—according to a transference predisposition (Racker, 1968). Thus, the transference in all its diverse expressions is the result of both factors. Transference phenomena need not be verbally "exported" into the relationship to affect its functioning.

Analogously, in the consultant there is the countertransference predisposition and the present real experience between client and consultant. The two together result in the countertransference. In the narrowest definition, countertransference derives from a specific reaction to an individual's transference. A more useful definition includes all the feelings and attitudes toward a client that are derived from earlier situations in the consultant's life and displaced onto the client. Some aspects of the countertransference are specifically reactive to the client, but other aspects are independent of the particular client (Natterson, 1991). This broader definition reflects the consultant's own unconscious reaction to the client, although some aspects may be conscious (Moore & Fine, 1990).

Racker (1968) expands on this idea by noting that it is the intention of the consultant to understand the client. Hence, he has the predisposition to identify with the client as a means of understanding. The extent to which the consultant is consciously in tune with the various aspects of the client's personality is the degree to which the consultant can empathize with the client. This process is called "concordant identification." Each part of the consultant's personality identifies concordantly with the corresponding part of the client's personality (ego with ego, etc.). To the extent that this process fails and the consultant unconsciously rejects these identifications, complementary identifications, in turn, become intensified, with the result that the consultant becomes the embodiment of the client's internal objects.

As highlighted in the example, countertransference does not simply designate a specific transference in the consultant to the client deriving from the past unresolved unconscious conflicts of the consultant. Rather, it is more complex and subtle, inseparable from the character, values, attitudes, experiences, and interaction of the consultant. If fully appreciated, it can then serve as a basis for the empathic understanding of the client by the consultant.

Regression in the service of the consultation, while essential to the understanding of the issues at hand, also leaves the consultant particularly vulnerable to the effects of projective identification. While projective identification is hypothesized to be active in the earliest periods of life (Klein, 1985) and to be reactivated in intense dyadic (Horwitz, 1985) and group situations (Bion, 1961a), little has been written about its important role in the consultative relationship.

As the consultation continued, the owner became aware, through the consultant's report, that issues of succession and the role of his children needed to be addressed. This activated an intense anxiety in the owner over his own mortality. The owner, through the use of projective identification, rid himself of his own anger and depression over this issue. The consultant, who had been performing his job well, introjects the owners' rage, suddenly experiences the "numbing sense of reality," and feels de-skilled. Simultaneously, the patriarch's projection leads to persecutory anxiety concerning retaliation by the external object [the consultant]. The patriarch felt hurt and vulnerable by having his omnipotence questioned, and feels a concomitant need to control and dominate the object. He also experiences increased paranoia, and so attempts to manipulate and control the consultant in order to minimize the possibility of retaliation. He does this by rejecting the consultant and asking for a more senior consultant to review the consultation. By so doing he seeks to repair the hurt already experienced and to avoid further intrapsychic damage.

The process of projective identification, as can be seen by the example, is not solely confined to the realm of the pathological. Because this process has both a subject and object, it serves as "a central bridging concept, linking our understanding of the psychological functioning of the individual with the transpersonal functioning of human systems" (Shapiro & Carr, 1991, p. 21).

Due to the pervasive nature of projective identification in working with a family business, it is important to more clearly define this concept. From the intrapsychic standpoint, Horwitz (1985) points out that the defense of projective identification differs from pure projection in that it derives from a variety of motivations in addition to the desire to rid the self of unacceptable impulses. He notes that wishes to dominate, devalue, and control—based on primitive envy and the wish to cling parasitically or re-fuse to the valued object—are among the other motives. This aspect is clearly seen in the example by the owner's attempt to diminish the consultant and direct the consultation. These projections, however, also lead to persecutory anxieties concerning retaliation by the external object, which heightens the subject's need to control and dominate the object (Horwitz, 1985).

From the interpersonal aspect of projective identification, it is imperative to recognize the effect of the projected material on the external object. The target of the projected content undergoes a fusion of identification with that material and its unconscious meanings and thus has the experience of being manipulated into a particular role (Horwitz, 1985). Bion (1961b) first delineated this as a cornerstone of group dynamics, noting that the group therapist who is the target of the projective identification of group members has the sensation of playing a part in someone else's fantasy. He or she experiences a "numbing feeling of reality" that makes it difficult to interpret the group's interactions. Bion goes on to caution that the group therapists must be alert to the occurrence of this feeling within themselves, be able to distance themselves from it, and then to rely on their affective experience to provide the source of their interpretations. The consultant working with the family business can be especially vulnerable to the vicissitudes of projective identification, and Bion's cautions apply.

The issue then arises as to what factors mediate the extent to which the external object becomes possessed by, controlled by, and identified with the projected parts. Horwitz (1985) suggests that projective identification grows only in the soil of "intimacy and intense involvement" (p. 22). The relevant variable that mediates the process of projective identification is the permeability of ego boundaries. Temporary regressive actions in the context of a healthy ego structure are recognized to be important aspects of psychic functioning. Such actions manifest as creativity, empathy, humor, and intimacy.

Consultants to the family business temporarily suspend their own ego boundaries in the interest of truly understanding the system. Such temporary permeability of ego boundaries enhances empathy and intimacy and leads to an enriched relationship. This suspension also increases the consultant's susceptibility to the vicissitudes of the projective identifications of the client, and increases the likelihood that the consultant's behavior will be modified by those projections. Family businesses, because they in many ways recapitulate the family itself, have a valence toward more permeable ego boundaries among members themselves, and hence a propensity to utilize projective identification defensively. On the other hand, the consultant to the family business resonates with the family structure in the business and experiences a particularly intense pull to regress, beyond the so-called regression in the service of the consultation. Only by developing a strategy of utilizing inner dialogue to arrive at an interpretative stance can the dangers of Bion's (1961a) "numbing sense of reality" be avoided.

Just as transference on the part of the subject can elicit countertransference on the part of the object, Grinberg (1979) suggests that projective identification can, in turn, elicit projective counteridentification. In the case of projective counteridentification, the consultant's reactions are independent of his or her idiosyncratic makeup. Rather, they derive solely from the intensity of the subject's violent projections and result in the subject's wished-for response from the object. The consultant then may experience the impression of no longer being his or her own person and of unavoidably becoming transformed into the object that the client unconsciously wished the consultant to be. The consultant may also experience those affects the client forced onto him or her (Grinberg, 1979). This was the case in the example when the consultant felt like one of the owner's children.

Thus, just as a particular transferential attitude can elicit a countertransferential response, so, too, a particular projective identification will evoke a specific projective counteridentification. The degree to which the consultant is susceptible to the effects of projective counteridentification is conversely dependent on his or her ability to assume an interpretative stance.

By understanding these processes, the consultant can use feelings as a lens to comprehend the experiences of the owner as he grapples with issues of succession. Using this knowledge to attain the interpretative stance, the consultant is in a position to help the owner overcome his reluctance to contemplate the future. The consultant is able to do this through management of the boundaries between self and client.

Boundary management is a set of mechanisms that individuals or groups employ to balance their need to protect internal reality with their need to interface with the external world. "In classical systems theory the organization imports resources and information across its boundary. The boundary separates the outer world or opportunities and challenges from the inner world of work and transformation" (Hirschhorn, 1990, p. 29). Significant to family business consultation are emotional and subjective boundaries. Miller and Rice (1967) note that when people take risks or feel threatened they set up psychological boundaries that challenge the functional pragmatic boundaries of the organization. The internally generated, often covert boundaries are as influential to the intervention as the external, overt ones. The role of the consultant often involves the particular management of the former. Due to the predilection of family firms to utilize projective defenses, which by definition blur boundaries, the consultant must utilize inner dialogue to clarify and then manage these internal boundaries.

## DISCUSSION

Some of the difficulty in addressing the issue of affective response on the part of a consultant derives from the lack of a clearly defined conceptual framework. Shapiro and Carr (1991) demonstrate that a "negotiated interpretative stance [based on an inner dialogue] provides us with a way of engaging and making sense of organizational complexity and our own place and worth as individuals within it" (p. 6).

One of the authors of this chapter, in previous research on the consultation relationship (Rosenthal, 1985), found that the perceived role of the consultant influences the client's perception of the quality of the consultation relationship. The client's perception of the consultant is a result of many variables, including experience with other consultants, internal organizational issues, the symbolic meaning the consultant represents for the client, and the personalities of the individuals involved. These variables are representational in character; they operate in the realms of reality and fantasy within the boundaries of the relationship. If we also take into account the unique nature of the family business (McCollum, 1990), and how its covert life resonates with family concerns, it appears to be all the more likely that the consultant will be perceived as an outsider and serve as the repository of unconscious projections.

In any consultation, the consultant serves as a vehicle to be used by the client. Herein lies the importance of transference and projective phenomena as they have an impact on the ongoing consultation relationship. Referring to the role of the family therapist and countertransference, but clearly implying the broader range of projective phenomena as well, Scharff and Scharff (1991) note:

> Countertransference, in our view, refers to the totality of the affective responses that occur whenever the couple or family creates an impact that penetrates beyond the therapist's conscious and relatively reasonable capacity to understand, beyond the central self. When this happens, the family or couple's object relations reaches an area of the therapist's unconscious. Training and personal therapy prepare the therapist's psyche as fertile ground in which these internal experiences can take hold, in which the growth of meaning amidst uncertainties can be cultivated, out of which is garnered a harvest of intimate understanding of the family from the ripened countertransferences. In this way, therapists allow themselves to be substrate for newly emerging understanding.... (p. 66)

As this is true for the family therapist, so it is for the family business consultant. Though the consulting relationship is not primarily therapeutic in nature, the relationship can elicit strong feelings from both parties. Given the powerful resonances of family life inherent in family business, the unreflective consultant can be seduced into taking on roles that serve the covert emotional needs of the client or self rather than overt requirements of the business.

The consultant's ability to reflectively acknowledge the interplay of fantasy and reality, and, in the best case, to teach the client to do the same, is the key to unlocking the productive insights offered by transference and projective phenomena. This ability allows both parties to avoid sinking in the mire of the client's projected "numbing sense of reality" (Bion, 1961a).

The consultant must recognize that in actuality one is not neutral, but, instead, one affects the system merely by entering and observing it. At best, the consultant has a history and an agenda that is an inseparable part of what is brought to the consultation. The consultant must try to consciously separate this personal agenda from the client's.

The consultant's past is *always* present in his work. Even those consultants with little background in working with family businesses can be susceptible to resonating with the family dynamics of such businesses. Various family client members may remind the consultant of past experience in business or as a family member. The consultant must be alert to noticing such powerfully seductive resonances and thereby avoid regressions. Those who do not reflect on their affective experiences often become "lost in familiar places" (Shapiro & Carr, 1991). In other words, they might find themselves confused and disoriented by their feelings or making inappropriate interventions even though they are engaged in a familiar activity (i.e., consultation).

Whenever consultants enter into a new consultation situation, they must "clear the screen" of perception and focus on the present situation. The consultant must seek to understand "what belongs to whom"—in other words, to differentiate countertransferential phenomena from projective phenomena. Clarifying the origins of such contributions in the present situation begins to establish a road map for the consultant, so he or she can better navigate those familiar regions.

At times the consultant has to fight back the impulse to respond to the client as if he or she were a family member of the consultant. The underlying dynamic, amplified by the familylike resonances of the family business, that prompts this countertransferential impulse to distort the roles must be attended to closely by the consultant. For example, when consultants who are trained to be neutral complain about their client's characteristics, his or her behavior toward the consultant, or his or her mannerisms, it is likely due to unexamined countertransference or projective counteridentification in response to specific elements of the client's behaviors.

There may be times when the consultant does not succeed in sorting all this out within himself or herself before viewing the client as something other than who he or she really is (Money-Kyrle, 1988). "If we were omniscient [consultants], the only countertransference we should experience would be that belonging to those intuitive periods when all is going well.... Yet it is precisely in them [the difficult periods], I think, that the [consultant], by silently [analyzing] his own reactions, can increase his insight, decrease his difficulties, and learn more about his [client]" (Money-Kyrle, 1988, p. 31). It is through this process of entertaining an inner dialogue that the consultant can achieve an interpretative stance and thereby be of assistance to the client. It may be added that in so doing, the consultant can also learn a great deal about herself or himself as well. The consultant has to take an internal reading of himself or herself and the work that he or she is doing. The consultant has to understand what the client is doing to him or her that is promoting the confusion, and the consultant has to see it in a developmental context.

Such a view entails separating reactivated personal developmental issues within the consultant from the unresolved developmental issues that influence decision making by the leadership. As Kets de Vries and Miller (1984) point out, the understanding of such pernicious developmental issues within the company is critical. Not only must full recognition be given to the separation of real and symbolic roles that executives fulfill in a family business, but it is equally important for the consultant to recognize those symbolic roles which he or she fulfills for the leaders and vice versa.

As was noted earlier, the consultant is not a neutral bystander but rather a participant observer in relation to the client system. At the boundary where the two meet, they join or pair when working together. The consultant has to face the challenge of becoming an insider while retaining the qualities he or she brought from outside. This implies a capacity for managing boundaries that will be particularly necessary to serve the "containing" function of the system.

## SUMMARY

The challenge for family business consultants: How are they able to differentiate between what is a reaction to present actions and what more appropriately belongs to the past? In particular, family business consultants enter into a system which, because of its special nature as a narcissistic extension of the founder, can be especially confusing.

Such a system is prone to utilize relatively primitive but powerful defenses, such as projective identification, when it encounters a perceived threat. Furthermore, transferential responses are likely to be compelling because they are reinforced on a daily basis due to the familial relationships of the managers. For the consultant, the similarity of the dynamics of family-owned businesses to family dynamics provokes in the consultant resonances to his or her own family.

As the consultant attempts to be both empathetic and understanding of the situation, the echoes of personal family dynamics are likely to result in some permeability of ego boundaries. If, as in the example, the owner relates to the consultant as the good son, what is more natural than for the consultant to respond and relate to the owner as the benevolent and giving father? The consultant is thus primed to experience countertransferential feelings and react with his or her own primitive defenses, including projective counteridentification.

Add to this the notion that these defenses are not simply intrapsychic, but are interpersonal as well. The utilization of projective identification and projective counteridentification implies an object as well as a subject. When such defenses are utilized, the aim is not only to eliminate undesired impulses, but also to control those impulses through reintrojection (i.e., identification with the object). The object of these defenses is thus prone to experiencing the numbing sense of the reality that is not his or her own, but rather is a product of the subject's unconscious, and to become lost in an affective response to what otherwise would be a comfortable and familiar place (i.e., the consultation).

How does the consultant manage to avoid such compelling pitfalls? Furthermore, how do consultants avoid acting on their own defensive need to manipulate the system and control their own anxiety as a result of being inundated with these externally generated impulses? The family business consultant can take no comfort in the construct of therapeutic neutrality. The consultant instead serves as the container of the feelings, hopes, wishes, and dreams of the client. Unless this role as container is acknowledged and accepted, consciously by the consultant and unconsciously by the client system, there is a significant profitability that the consultation will not be successful.

Unlike the psychoanalyst, the consultant must do without the traditional luxuries of the consulting room or therapeutic space with the concomitant ability to manage boundaries and control that space. Rather, the family business consultant must metaphorically create a therapeutic space from within which he or she can collaborate with the client. The key to the creation of such a space lies in the ability for the consultant to entertain an inner dialogue. Both the consultant's own feelings and a thorough understanding of how those feelings can be generated must inform this dialogue. Through this dialogue the consultant can avoid acting on those echoes of personal family life provoked by the family business client, and, instead, can achieve an interpretative stance from which to make sense of the system that is presented to him or her.

On the other hand, if the consultant is unaware of the pervasive effects of transference, countertransference, projective identification, and projective counteridentification, he or she is likely to be unable to entertain the inner dialogue. The result will be that the consultation will fail.

In order to minimize the possibilities of failure, a variety of strategies are available to the consultant to help promote an inner dialogue. Self-awareness of excessive feelings (or lack of feelings) arising in the context of a consultation relationship can serve as a signal that transference distortion may be operative. Use of a shadow consultant, who is one step removed from the consultation, can serve to separate internal from external realities. A supervisor can aid in the analysis of the consultant's feelings to assess whether they are either too excessive or too well defended against. And, at the very least, regular reality checks with the client can ensure that the consultation is on course and can help prevent the acting out of inappropriate feelings.

As demonstrated by the case study, the consultant-client relationship exists on a number of levels: professional, interpersonal, class, race, gender, age, values, insider-outsider, fact-feeling, past-present, and fantasy-reality. These multiple relationships are always present on both sides of the interaction. How the boundary is managed by both parties will be, in part, a function of the consultant's ability to achieve the interpretative stance. Such a stance can minimize the negative influences of intrapsychic processes on good leadership practices in the family-owned business. Effective boundary management will, in turn, determine the success of the relationship and the quality of the consultation.

## CRITICAL UPDATE OF CASE CONSULTATION

Two years have passed since the consultation. The owner remains to this day unable to face the implications of the consultation with regard to his own succession and mortality. Rather than act on the substance of the consultation, the owner has chosen to embrace only the surface recommendations. As a result, a family board was created, a senior vice president for human resources was hired, and a formal executive team was formed. The owner will not name a successor, and no succession planning has taken place.

The owner continues to run the Company as a benevolent despot. The executive committee serves at the pleasure of the owner. Often intimidated, they have to work around the founder to effectively do their jobs. The family continues to fight, and the Company continues to react to lost marketshare. In addition, perhaps as a reaction to the intrapsychic implications of the consultation, as well as to the dynamics of the marketplace, another consultation firm has been allowed to do reengineering without addressing issues of family and succession.

## REFERENCES

Bion, W. R. (1961a). *Experiences in groups.* New York: Basic Books.
Bion, W. R. (1961b). *Learning from experience.* London: Karnac Books.
Grinberg, L. (1979). Countertransference and projective counterprojective identification. *Contemporary Psychoanalysis, 15*(2), 226–247.
Hirschhorn, L. (1990). *The workplace within: The psychodynamics of organizational life.* Cambridge: MIT Press.
Horwitz, L. (1985). Projective identification in dyads and groups. In A. D. Colman & M. H. Geller (Eds.), *Group relations reader 2* (pp. 21–35). Washington, DC: A. K. Rice Institute.
Katz, D., & Kahn, R. (1971). Open systems. In J. G. Maurer (Ed.), *Readings in organizational theory: Open systems approaches* (pp. 13–29). New York: Random House.
Kets de Vries, M. F. R., & Miller, D. (1984). *The neurotic organization.* San Francisco: Jossey-Bass.
Klein, M. (1985). Our adult world and its roots in infancy. In A. D. Colman & M. H. Geller (Eds.), *Group relations reader 2* (pp. 5–19). Washington, DC: A. K. Rice Institute.
McCollum, M. (1990). Problems and prospects in clinical research on family firms. *Family Business Review, 3*(3), 245–261.
Miller, E. J., & Rice, A. K. (1967). *Systems of organization.* London: Tavistock.
Money-Kryle, R. (1988). Counter-transference and some of its deviations. In E. B. Spillius (Ed.), *Melanie Klein today: Vol. 2. Mainly practice* (pp. 22–33). London: Routledge.
Moore, B. E., & Fine, B. D. (1990). *Psychoanalytic terms and concepts.* New Haven, CT: Yale University Press.
Natterson, J. (1991). *Beyond countertransference.* Northvale, NJ: Aronson.

Racker, H. (1968). *Transference and countertransference.* New York: International Universities Press.

Rosenthal, J. S. (1985). The interpersonal aspects and structural dynamics of the consultation relationship: An exploratory study. Unpublished dissertation, Boston University.

Scharff, D., & Scharff, J. (1991). *Object relations couple therapy.* Northvale, NJ: Aronson.

Shapiro, E., & Carr, W. (1991). *Lost in familiar places.* New Haven, CT: Yale University Press.

Sullivan, H. S. (1953). *The interpersonal theory of psychiatry.* New York: Norton.

# 7

# Understanding the Impact of the Founder's Legacy in Current Organizational Behavior

*Alexander H. Smith, Ed.D., A.B.P.P.*

From an organizational perspective, a good day at the office is not that different from a good day in the monastery or convent. Organizational processes in religious communities resemble those in business and health care systems. The patterns of established group process, collective decision making, task maintenance, affective relationships, and historical growth and development all converge around a commitment to an ideal that the group maintains. Businesses and religious communities share in an overarching commonality of organizational behavior: They are tied to the founder's vision and legacy. In this way the aims of health care, business, and those of religious communities are very similar in their manifestation as group behavior. The examination of organizational consultation to religious orders in this context illuminates a different perspective of the same process, one born primarily of a very long historical development within a contemporary context. A religious community successfully maintains its continuity in ways very similar to those found in corporations. Members perceive and interpret their founder's purpose and ideals, and internalize them within the complexity of both consciously formed attitudes and unconscious fantasies.

This chapter considers a contextual analysis of organizational forces as they have been observed in Roman Catholic religious communities. These include the historical setting of the order's development, the continuation of the founder's vision, the cultural transmission of these ideals through the group's history, and the tensions between the current group process and the original ideas of the community. Consultation approaches with similar organizational processes appear in comprehensive analyses of social systems and their defensive functions (Levinson, 1991). These methodologies include systemwide problems in nursing-service practices (Menzies, 1984), studies of decay of organizational ideals in management decision making at the National Aeronautics and Space Administration (NASA; Schwartz, 1991), and shifts of power in major corporations (Zaleznik, 1984). Applications of organizational consultation have been made in resolving splits in social perception of personnel at an oil refinery (Hirschorn & Young, 1991), and the psychodynamic management of upheaval in mergers and acquisitions (Gilkey, 1991). Likewise, Caplan and Caplan (1993) have applied mental health consultation to the organizational level in work with groups of ministers and parishes as well as to more comprehensive efforts in mediating between community factions in wartorn communities in the Mideast.

## THE IMPACT OF VATICAN II UPON ORGANIZATIONAL RESTRUCTURING

The parallels of organizational change in business, seen in problems with new management, shifts in organizational goals and missions, and the rethinking of the visions of a group's founder, are easily recognized in the effects of Vatican II upon religious orders. Along these lines, contemporary readers may wish to recall the historical significance of Vatican II upon the Roman Catholic Church in the 1960s. Vatican II mandated that religious orders modernize themselves within their founders' original ideals. These changes addressed a range of traditions and philosophical positions, from dress and living arrangements, to training of personnel, to renewing governing structures. For many communities these times of change were ambivalently welcomed and organizationally confusing. Often these years of transition were quite painful, as most communities lost large numbers of members, frequently as many as half over the next 2 decades. Some communities survived amazingly well, but not many. As one brother put it, "you just had to be there to appreciate what it was like . . . everything was falling down all around us . . . guys you had spent years with, gone the next day . . . chapel becoming emptier by the month . . . you didn't know which friend was next to leave."

Most communities have undergone some kind of emotional upheaval since the late 1960s. For many religious communities change was like taking the cap off an overheated radiator. Emotional flooding of previously dammed-up resentment toward authority, and the need to grow individually in ways interdependent with the order, often influenced the newly formed self-government in groups where there had been an overcontrolled, oppressive leadership. Group decision-making practices that attempted to democratize the community's leadership were often in reaction to the previous authoritarian styles. The understandable pressure of these times drove communities to change, but from an organizational perspective they may not have been as successfully democratic as intended.

Kernberg's (1985) distinction about participative democratic processes in a group illustrates this point: Those that are founded by an imposing larger majority are not necessarily truly democratic. Symbolic changes, like the rejection of wearing religious garb or of undertaking penitential practices, were often the result of a "majority rules" approach contextualized by the revolutionary and reactive "authority = abuse" era. To some members these methods were ephemeral and naive, to many they were mutinous, to others they were a breath of fresh air. Invariably, toes were stepped on, and the residual effects often remained split-off. The consultant can unwittingly step on toes again by not attending to the phenomenology of the current context—hence the need for a careful organizational diagnosis.

A simple example is that of a community's decision to modernize its chapel for its house of studies. The original chapel had carved choir stalls and frescoes. During the early post-Vatican II period, the community had decided to "hold the line" with major changes and renovation. But a later split within the community had resulted in an administration that advocated for rapid change. The original chapel was gutted; the choir stalls were replaced with simple chairs in a circle. The Liturgy of the Hours was reduced to a fraction of its original observance. The rest of the community was expected to follow suit. Years later, the chapel became the metaphor of the community's struggle. One member who had survived "the holocaust," as it was now termed, simply stated, "All the guys who came in and ripped out the original chapel are now gone . . . some of us had our hearts ripped out with it" (Smith, 1997).

## IDENTIFYING THE CURRENT GROUP'S CONNECTION TO THE FOUNDER'S IDEALS

The consultant, just as in a corporate setting, must be attuned to the connection each member makes to the founder's ideals, and, conversely, must appreciate the hopes that the entire community's ideals hold for its members (Levinson, 1991). He or she is uniquely positioned to observe and to interpret what this community gravitates toward and how it maintains its commitments to its founder's ideals. The consultant needs to become immersed in the "feel" of the founder's original and contemporary influences. Every member has some reason, both spiritual and psychological, for having chosen this particular community. These motivational systems comprise the "glue" of the group's cohesion.

The group's "manifest" cohesion is observed in the attitudes and behavior common to both the individual's identification with the founder and the group's enculturation of the founder's ideals in rituals, observances of custom, dress, and philosophical and theological outlooks. These ideals provide personal satisfaction and meaning to its members as they find themselves sharing similar values, later identified with the founder's ideals. Yet considerable tension can develop within a community around the common interpretation of these aims.

An example of a commonly occurring tension at the group's boundary between its contemporary cultural context and its founder's ideals is work-related earnings outside the community. Most orders have had to begin supporting themselves. Those members working in professional settings, who now seek parity in compensation for teaching, school administration, or health care, can command larger salaries than before. This can create an unwieldy power differential for groups that have not worked through the necessary structures. Those members who contribute large amounts of money may feel unfairly taxed by those who are less active. This can be a thorny problem for the leadership, and can create a subsystem of difficulties in organizational boundary management, given the matrix of leadership effectiveness and/or basic group assumptions.

One community had several sisters who originally were trained as nurses when the nursing role was one subservient to physicians. As nursing began to seek a more independent and consultative role in relation to mostly male physicians, these same nuns, influenced by the need for changes in their status as nurses, obtained master's degrees and commanded substantial salaries. The order, holding to its traditional interpretation of the vow of poverty, expected these salaries to be promptly handed over to the organization. The nuns were expected to be obedient and silent about the glaring imbalance in members' productivity, much like the physicians expected them to take orders. The unexpressed anger and unresolved entitlement became cancerous to the group as a whole. From an organizational process perspective, the leadership had not been able to mediate the aims of the order within its cultural context by addressing it directly with the membership, and thus helping the group evolve more adaptive expectations. It was nearly ruinous to let the group split into dependent and counterdependent factions, as "traditional" and "new." These maneuvers were the outcome of a regressive passivity on the part of the leadership.

The leadership's interpretation of the founder's ideals thus becomes something of a crucible in which the community's ongoing group process forges a continuity, much like upper management executives must continue to be clear in fulfilling the "mission" of the company founder and its current board of directors. Projective processes are quite observable, in that each individual "fills in" his or her unique understanding as part of a consensus building. The usual issues of authority, power distribution, leadership, alliances and subgroupings, and maintenance of group boundaries (Klein & Gould, 1973) all are cast within the historical context of the community's history and development. The consultant can use the community's rules as an inkblot against which its members project their own unresolved conflicts, longings and demands, and help people understand that system issues produce such effects, not individual dynamics.

## THE COMMUNITY'S "CHARISM" BRIDGES THE FOUNDER'S IDEALS AND THE CONTEMPORARY SOCIETY

Just as corporations such as Apple and Microsoft live out the influences and visionary focus of Steve Jobs and Bill Gates, respectively, religious communities manifest a continuity of their founders' ideals. Depending on its current leadership, any significant historical change in the community's boundaries usually occurs in relation to the larger culture and Church. The "personality" of the community, like that of a major corporation, is often what attracts and keeps members. It establishes both a boundary and its permeability to the outside world. It is the avenue through which new members enter and become enculturated to the group's ideals and ultimately to the founder's purposes. In Church parlance this quality is often referred to as the "charism" or dynamism of the particular community, through which it expresses its purposes, unique tasks, and missions. Religious communities consciously maintain these expressive styles unique to their own traditions and usually oriented toward the founder's personality. Many communities pride themselves on carrying out not only the founder's ideals, but also imparting his or her affective qualities in a positive way. Franciscans have traditionally focused on the joy and freedom of St. Francis, Jesuits carry forward the zeal and keen intellect of St. Ignatius, and Benedictines are known for their hospitality. These emphasized qualities are also driven by the complexities of the group's tradition and its identification with the founder. They comprise part of the order's methods of task maintenance and group continuity. They are a kind of "school spirit" or "espirit de corps" unique to each community. This spirit is more than a "persona" of the community, as it ties each person to the founder's vision in a personal way, eliciting that member's personal interpretation and commitment to the group. It links the heart of the individual to that of the founder.

Every community, whether more recently founded or dating back several centuries, has maintained its tasks and attempted to adhere to goals in accordance with the founder's vision or mission, as set forth in the rule of the particular congregation. These historical developments within the order are a template of a group unconscious created by the founder's interaction with the initial followers. The contemporary community inherits several generations of interpretations of group tensions in relation to the founder's ideals and methods. The consultation process can easily miss the subtleties of this historical context by which systems of attitude, motivation, and decision making have been passed on to successive generations.

For example, the Benedictine tradition has lasted since St. Benedict founded the order in 580 CE. This phenomenon is itself a remarkable testimony to its rule and self-regulation (Rippinger, 1990). However, there have been many reformations that occurred within the greater contexts of changes in the Church, and in the politics of each locale. Political climates greatly affected their evolution. Church and state were not separate in the way we are familiar with them. The interest of royalty greatly affected each abbey's prosperity as well as the influx of personnel. The social perception of the monk or nun evoked both positive and, later, negative projections of kings and queens who could make or break the abbey's coffers. The unexpected opulence of Cluny led to a reformation by Stephen Harding and Robert Molesme at Cîteaux in 1098, aimed at the restoration of Benedictine (now called Cistercian) ideals. Several centuries later (1664), we find the Cistercian order itself going through a very similar process in which a new abbey is established, with the strictest regulations and penitential atmosphere, at La Trappe. Similar changes are easily identified in other traditions. The Franciscan order of men has undergone three major reformations as well as several minor ones. The Carmelite nuns and friars, under Teresa of Avila and John of the Cross, experienced the "Spanish" reformation of the order in the 16th century.

Adaptation and change mark religious order development within the Church's history. Newer communities were shaped by cultural demands and historical contexts of the Church during the Counter-Reformation. Vows of poverty, chastity, and obedience were called "simple" rather than "solemn," and the moral weight attached to their observance shifted. Each one had a particular "charism" or special motif to its calling, most of which in some way concerned the expulsion of heresy and the protection of Church doctrine. Some orders were missionaries, while others worked with the sick and poor. The Jesuits, for example, were viewed as the Church's intellectual commandos sent to fight the heresies. Virtually every order, congregation, and society was subject to similar vicissitudes as it emerged within its original cultural setting and passed through the matrix of historical pressures. These historical determinants formed a major impact upon contemporary group process and the continuity of efforts to live out the founder's goals.

## DISCOVERING THE CONSULTATIVE THEME: THE IMPACT OF THE FOUNDER'S IDEALS UPON CURRENT GROUP ASSUMPTIONS AND LEADERSHIP STYLES

Members of religious communities easily form an "assumption" (Bion, 1961) about the group that reflects a collectively held longing for continuity, permanence, and protection. The group and its leadership are imbued with a quality of omnipotence and perpetuity. When this occurs, there are probably several members who have patterned around an idealization of the founder, often as a defense against commonly experienced abandonment fears and split-off merger longings. These developmental arrests, as they become part of a community-shared unconscious, push its members to behave in regressively demanding or passive ways. This is particularly true during times of transitions in leadership, financial crises, or scandals about the behavior of one of its members. It is as though the realities of human nature are unacceptable to this assumption of "larger than life" myth that the membership has set forth about itself. The membership in these times has projected upon the order the ongoing demand of being bathed in its founder's strength and glory. Though the founder is not usually present, the community's leadership is. The group behaves as if there were an all-knowing, omnipotent quality to the founder that has been passed onto the group leaders like royal lineage. There may be a split at these times between the membership's feeling about its founder and those directed toward current leadership, much like naval personnel on a ship will maintain their idealization of the captain and displace their anger to the executive officer (Kets de Vries & Miller, 1991). The result can temporarily paralyze the group's ability to remain task oriented, and to face the exigencies of problem-solving behavior.

For example, one community had been used to large institutional settings where several of its "second-class" membership (laybrothers) would cook its meals, and the remaining clerics would simply show up to eat. In the years following Vatican II, this community, like so many others, lost a considerable percentage of its membership, including its laybrothers. The large institutions began to close, and the remaining communities had to fend for themselves at mealtime. The rule had been developed in an era when the local custom was to give alms to religious, reinforcing the notion of being passively fed. In later centuries the order had become used to a more sophisticated means of support whereby larger endowments and donations were made to the central office and each large house would partake of a fairly regular beneficence. Again, the laybrothers (as opposed to the priests, clerics, and seminarians) would be required to procure the food and prepare it. The unconscious expectation was based upon a notion handed down through literally hundreds of years—that "God will provide." Now the community houses were faced with the problem of how to support themselves, as many of these resources had dried up since the "golden era." Members struggled not only with the fact that their meals were no longer "fed" to them, but also with the realities of making a living. The group as a whole had lived with an unconscious legacy that, because it had established itself as an extension of the founder's ideals, it would never want for supplies. In its dependency upon the order (and fantasized merger with the founder), many in the group had come to view the order as invulnerable to hunger. It was as if by merging with the ideals of the founder one could expect that the order would survive any kind of stressful change, and still maintain the "original" institutional mode of living.

The order's perception of the laybrother contained a central issue. The clerical component of the order, as opposed to the laybrothers, had both split-off its demands toward the leadership to be fed, both literally and metaphorically, and projected its anger about the leadership onto the laybrothers. The "second-class" menial status assigned to the laybrothers provided the clerics further opportunity to defensively avoid confronting the leadership, thus protecting the idealization defense historically maintained. The laybrothers played out the group's projected devaluation, sleeping in poorly ventilated, common quarters while the priests had their own rooms. The laybrothers kept their anger to themselves, rationalizing it, with the help of the clerics, as suffering to be "offered up." Some priests in the community even produced quasi-theological documents advocating this caste system.

The development of the distinction of laybrother appeared to have served at least one unconscious motive in the order's history. This resolution was a commonly occurring theme in the founding of many religious communities—and corporations. It attempted to preserve the clerics' special relationship with the founder. The order was now faced with the painful recognition that this decision had been a regressively motivated one that actually was not in keeping with the founder's ideals at all but with the unresolved demands of his contemporary followers. The opportunity to examine the effects of this very old group assumption—"I am more special to the founder as a cleric"—came as a shock to many who had themselves played out these roles, both as clerics and as laybrothers. Narcissistic leaders had seized upon this distinction and used it to maintain a pipeline of scapegoats through generations within the order, thus further splitting the community and avoiding questions of effective leadership.

## RECOGNIZING REGRESSIVE LEADERSHIP AND GROUP PROCESS IN OBSESSIVE-PARANOID, DEPRESSIVE-IMPULSIVE, AND NARCISSISTIC STYLES OF COMMUNITY

Communities attract personalities by their mission. The mission itself of each community can affect leadership styles, which vary idiosyncratically across individuals and groups. Organizational processes, however, can also mirror individual psychopathological counterparts from a perspective of group members in relation to leadership, and from a complementarity in leadership style (Levinson, 1991). Prevaling group assumptions will attract members with similar values, satisfying both individual and group needs.

Three major assumptions are identified here for heuristic purposes. They reflect object relations, self psychology, and general systems theory influences. A complete typology is beyond our purposes, but these three may more or less integrate the major processes and characteristics and help the consultant with the following tasks: (1) Identify the contextual elements occuring in a common group tension (Ezriel, 1952) in relation to the leadership. (2) Utilize the community's history and biography of ideals to provide a provisional matrix of diagnostic and intervention opportunities to the organizational consultant (Klein, 1992). The obsessive-paranoid, depressive-impulsive and narcissistic dimensions are considered here as indicating a mutually regulating system of the group in relation with its leadership and within the context of the founder's rule and personality characteristics (Kernberg, 1985). The group transferences often made to the consultant are considered concurrently.

## Obsessive-Paranoid Styles of Group Process

Just as corporate environments vary in their degrees of formality and the "degrees of freedom" within which individuality is tolerated, it will become apparent to the consultant that particular religious communities will exact a more detailed and careful observance than others. Such a community often comes across as overly controlled. It will retain a somewhat punitive tone to any deviation from its rule, customs, and traditions. The group is often overtly passive and compliant, but covertly seething. In its interpretation of the order's rule, leadership often expresses its own sadistic impulses through a "letter of the law" style that pushes for submission. Passive-aggressive maneuvers and/or psychosomatic problems among the membership are quite noticeable. Members who introduce newer ideas or fresh thinking are viewed with suspicion.

In stressful times the obsessive-compulsive features of the group give way to paranoid defenses that can reflect poor reality testing in both leadership and members. This defense serves to bind the group against further danger of fragmentation within the community, thus localizing a particular issue in the outside world. The projection of danger is usually onto an individual or group who could potentially disrupt the community's maintenance of order. The group's cognitive self-appraisal goes something like this: "We are secure when we follow the rule.... The more details we attend to, the safer we are.... The rule of the founder is our life-line.... We must protect ourselves from infiltration by those who would make us unfaithful to our rule." This should not be taken to mean that "traditional" communities, that is, ones wearing religious garb, are necessarily obsessive-paranoid, as they may enjoy a well-functioning and effective group process. "Modern" communities can be just as fearful of outsiders attempting to "change us back" and may set up similar group defensive patterns.

The consultant to the obsessive-paranoid dominated community is frequently viewed in a similar way, as a potential threat to the status quo, as one who will disrupt the traditions of the group.

Often such communities are the most reluctant to seek consultation. They will consistently test the consultant to see whether he or she will mirror the community's need for "sameness." Among leadership styles, there is usually less manipulation of the consultation, fewer attempts to undermine it, and greater responsivity to well-timed interventions that communicate an understanding of the group's need for control, order, and protection. As the leadership recognizes that the consultant will not attempt to take control, there can be a stronger "working alliance." This relational style is, however, less easily threatened than the narcissistic one discussed later.

## Depressive-Impulsive Styles of Group Process

The juxtaposition of depression and impulsivity may seem like an odd combination, but it actually reflects two dimensions of one process. In more sexist times we would have called this affect-dominated style "hysterical." In addition to the term being offensive to women (the term *hysteria* is from the Greek word meaning "womb"), it also fails to convey the scope of disinhibition and affect-driven behavior so characteristic of such groups. The assumption of the group is not based on fear, as in the obsessive-paranoid community, but upon a need for expressed behavior from others. In psychodynamic terms, the group searches for a lost ideal-mother imago that it shares as a common longing and projects the demand into its leadership. These communities are often motivated by strong wishes for affection and connection to others. They are cynically referred to as the "touchy-feely" communities.

There is often an ounce of truth in such a term, as one can observe consistent relation-seeking behavior within and across the membership. Not infrequently, the leadership is rather weak or indecisive. It does not or cannot address the emotionally charged issues from a decision-making perspective. Frequently, the leadership is bogged down in an "egalitarian" role, in which the leaders themselves, directly or indirectly, seek nurturance from their constituency. Often this behavior is rationalized in a solicitous way as being "open to others" or as "not being above others" in the community. No one will verbalize or recognize the nature of the community's separation from "the lost mother." The lack of strong emotional leadership sets the stage for rounds of depressive discouragement and for intermittently impulsive behavior that systemically defends the group against loss and abandonment. From an organizational perspective, this cycle may be viewed as, alternatively, retaining the unconscious fantasy of taking on (oral incorporation) and/or "breaking free" (counterdependency) from the needs of disappointing leaders. The group as a whole selects or utilizes members whose individual defenses are most vulnerable to disappointment in the leadership, and, through projective identification, expresses its disappointment or vicariously satisfies its longings for dependency by maintaining a collusion in the member's behavior.

The consultant easily becomes the target of dependency-based transferences. In contrast to the initial distancing suspiciousness of the obsessive-paranoid group, the consultant to a depressive-impulsive group will frequently discover in the initial stages a naive eagerness to jump into the consultative process along with an indiscriminate idealization of the consultant. The community may have cycled through several consultants. These groups will often pressure the consultant to use a special technique they heard about in the media. Not infrequently, out of desperation, the leadership will bring in consultants where there is no clear link to the task of the group, only later to de-authorize the consultant.

One religious order spent thousands of dollars on a training course in a mental development technique that was commercially available. Another spent a similar amount on testing its members for personality characteristics. The rationale for these decisions was never clear, except that better thinking and more concordant patterns of interest and stimulus preference would presumably lead to better group communication and effectiveness. Neither group ever addressed their respective organizational problems; their leadership decisions further weakened their already-compromised credibility, not to mention their strained budgets.

As the group has often felt unanchored by its leadership, in the initial phases of consultation there is likely to be an incorporation of the consultant's comments and behavior with an unconscious wish to replace the current leadership with that of the consultant. This pattern of boundary alteration is difficult for the consultant, as there are frequent "seductive" attempts to get the consultant to "act out" by assuming tasks that are structurally appropriate only to the group and its leadership. The group turns to the consultant and in effect says, "Please take care of us orphans. We can't do it ourselves." I am familiar with several instances where consultants have been drawn into leadership roles, paid as actual employees of communities, and designated with decision-making responsibilities, only later to be fired when the group did not find the consultant fulfilling its expectations.

## Narcissistic Styles of Group Process

Narcissistic/charismatic leadership styles consistently have the most destructive long-term effects on group cohesion, task maintenance, and flexibility of affective expression. A brief synopsis will refresh our understanding, and the observant consultant to religious communities will easily recognize the process. The central feature of the narcissistic leader is the projective identification of self needs (mirroring) into the group. The group is experienced as a container of self-esteem, or as a "self-object" (Kohut, 1977). The charismatic leader expresses or, more accurately, manipulates his or her need for admiration and security by appealing to the group's wishes for a hero. In the imago of invulnerability is an opportunity for endless mirroring from members who seek a strong hero. This collusive "trade-off" results in the group acting out a fantasized merger with the leader, making itself psychologically obedient to the leader. The difference between this style and the depressive-impulsive one is that the latter is driven by needs for nurturance and seeks these responses from the leadership or its substitutes, whereas in the narcissistic group the membership essentially complies with the demands of the leader in the fantasy of a merger.

The stages of group development never evolve beyond that of dependency upon leadership (Bennis & Shepherd, 1956) because the leader never permits individuation of members nor their interdependence on one another. The group's task focus is easily disrupted by the leader's needs, which seldom allow the fresh creativity that can occur as a group moves to its "working stages." The cognitive component is "I am one with the leader's strength or ideal and therefore invulnerable to abandonment and loss." The members typically incorporate the leader's perception of reality and never individuate beyond the controls that the leader imposes upon the group. Since the group can never completely satisfy the narcissistic leader, he or she typically needs a scapegoat to retain his or her own projected devaluations. These leaders often "set up" one after another of the members to assume that role. The group, in a reciprocal process, will select its next victim to be offered to the leadership's wrath, thereby maintaining its idealization of the leader. Individual members who recognize that "the king has no clothes on" are immediately treated with suspicion, if not actual exile. Usually subgrouping occurs, with the disillusioned, hurt, and angry members of the large group polarizing, often secretly plotting to overthrow the leader. The narcissistic leader often is so sensitized to any sign of disloyalty that he or she easily detects the counterdependence of the subgroup, thereby confirming the leader's paranoia and setting the stage for further rounds of scapegoating (Kernberg, 1985).

Religious orders are particularly vulnerable to narcissistic leadership. In communities where the founder's ideals have been interpreted as a mandate to inspire others, the system itself may contain a valency (Bion, 1961) for individuals with narcissistic qualities to more easily emerge as leaders. Religious groups easily adopt a dependent assumption of fusion or "primitive bliss." This maneuver often serves as a regressive social system defense against the group ever coming to terms with the very conflicts and issues that inspired the founder's efforts, as well as against the contemporary shared tensions and fears. Religious communities with narcissistic leadership characteristics are most difficult for the consultant. The consultant can expect to be drawn into a context of confusing and rapidly shifting boundary relations and to receive provocative and coercive projections. Often these groups have been languishing in institutional misery with little productivity. There are conflicting factions with highly charged issues and little sense of proportion or capacity to step back and self-reflect, and the consultant's interpretations are deflected like water on a duck's back, as they may be in industry (Kets de Vries and Miller, 1991).

The leadership will push the consultant to recognize certain issues as hopeless. This is usually to maintain scapegoating of individuals or subgroups through splitting and projective identification. The leadership has much to lose and little to gain by understanding or changing the effects of its behavior and attitudes; similarly, the group as a whole that supports this kind of messianic leadership bears the risk of losing the "golden fantasy" of fusion with its hero. The consultant is expected to do one better than the "messiah," and at the same time to never outshine him (or her). The leadership will be immediately suspicious of the consultant, and will frequently envy any actual progress made in the consultative process. Just at those points where actual change begins to occur, the leader will frequently undermine the consultant or even dismiss him or her for some "trumped up" reason.

Kernberg's (1985) typology of regression-prone leaders identifies one particular type who not only manipulates the members' love and admiration and keeps tight reins on the decision-making process, but easily sacrifices the group's value systems and/or organizational needs for political expediency. This "nice guy" approach is a glib one that basically involves avoiding the heat that comes with competent leadership. Consequently, such a leader adeptly turns conflicts among members of the community into ones that do not involve him or her. While there may not be the messianic grandiosity and control of admiration noted in more aggressive styles, the gratification is available to the leader in other ways, as in using the position for recognition beyond the community. Corporations, health care systems, and religious orders alike will have hidden scapegoating emerge where there has been a revolving door of personnel problems and turnover, often the result of pseudodemocratic leadership styles that require a scapegoat pipeline.

Consensus leadership styles include reaction formations to sadistic trends (Kernberg, 1985) that prompt the leader to be overly friendly to immediate personnel while severe conflicts, split-off by the leader and projectively identified into other individuals, persist in the outer layers of the group. Persistent problems of infighting located in specific houses can embody conflict for a whole province. The consultant might find the provincial superior initially affable and friendly, but continuing to keep the projective focus on the "problem-child" community. Comments like "I just don't understand why that one community in our order has such long-standing battles all the time. It seems like no matter who lives there, there is always a lot of trouble" usually suggest that the leadership is utilizing the scapegoated subgroup as an organizational defense. There is probably some inability to tolerate the derivative expressions of anger and disappointment that should be directed toward the leadership in such a way that creative resolutions can occur. A similar example of resolving scapegoating in an industrial setting in a Fortune 500 plant is described by Alderfer and Klein (1978).

In business cultures, a common mistake is to place a highly competent salesperson or technically skilled individual in a management position. Similarly, it frequently happens that a community will choose a leader because of some demonstrated competence in a particular area—one not related to management. The shared group fantasy is that because the leader has achieved competence in theological studies, for example, therefore he or she is equally competent in administration. This thinking within the group—"the quarterback will surely make a good class president"—in turn affects the individual who accepts such a position. Often out of the appeal to narcissistic needs, such an individual often struggles with the demands of the group as a whole and, because of a lack of adequate management skills, may come to avoid the difficult but necessary decision making of the position. This pattern of leadership selection in religious communities seems quite common. Again, these fantasies often seem to be formed around the unconscious group wish for an omnipotent leader and the assumption of dependency.

## ESTABLISHING THE FRAME FOR ORGANIZATIONAL CONSULTATION

As in virtually all types of organizational work, the consultant needs support from key personnel in leadership and administrative positions in order to develop a contract and formatting of service. It probably is better to decline consultation work if such support cannot be secured. When the elected administration is new and does not have the full support of its constituency, a commonly occurring pattern in initial phases of the consultation is that the leadership, often younger than many oldtimers, may retain only titular power of decision making, while the informal power of the community lies elsewhere. The consultant is wise to recognize, in the diagnostic stage, the real power brokers, even though maintaining a formalized, contractual relationship with the current administration.

Organizational consultation is not the same as mental health consultation, program development, or group psychotherapy (Klein, 1992). However, members of the community will easily confuse the two in response to the consultant, with the accompanying transferential distortions. When the contract is for organizational work, it is important that the consultant, particularly if he or she is a clinician, avoid getting pulled into a therapy role. When the community wants the consultant to address or resolve issues that are individually focused, the consultant should attempt to link this "resistance" to a larger focus that the group cannot sustain. At times this task can be difficult, indeed, as the community may be overwhelmed by issues that have never been adequately processed in a group format, ones that have been split off either by the members or leadership. The temptation is to give in to the group's pressure to address the isolated problem and to collude with the organization's defensive structures.

The consultant needs to keep the initial contract in mind at times when the group begins to expose its inevitable tensions while utilizing the individual problem as a hologram of the larger system tension. Many clients expect a modification of group psychotherapy. Some communities unconsciously associate the consultation with the antiquated monastic practice of "chapter of faults," in which members would identify one another in the group as having transgressed the rule and customs in some way. Religious communities can easily give a self-punitive spin to the consultative procedures, or even try to prompt the consultant to become the Grand Inquisitor. The consultant needs to teach the group how to go about discussing its identified problem areas and to demonstrate through the interpretative process that conflict can be faced adaptively. When these same projections occur later on, the consultant may wish to consider interpreting them within the larger fabric of the group issues. The consultant monitors boundaries between individual and group, between group and its leadership, and between the group as a whole and its relationship to larger social structures. The problem identified between the main group and its subgroups will have impact upon all these boundaries of the group.

## USING A "FOCAL CONFLICT MODEL" OF CONSULTATION

An adaptation of Whitaker and Lieberman's (1964) "focal conflict" model to organizational analysis is quite useful for sorting out the direction of "here-and-now" group tensions into "disturbing impulse," "reactive motive," and "current resolution." The balance of these deeper unconscious forces pertain to the type and content of the group's tension. They are often about dependency, anger toward authority, and wishes for autonomy and/or parity. The reformatting of this psychotherapy model into a consultation paradigm allows the consultant to ground his or her formulation in process or "ad hoc" data observed in the development of a particular theme. The manifest consultation event can thus be carefully tracked and quickly identified into segments or blocks of theme development; later the structural subgrouping may be identified. The consultant watches for how the group tries to express a wish, how it defends itself against the wish, and the style or manner of the solution in behavior and attitude. Often it is useful just to categorize the shifts or movements within the group according to these categories.

## A CASE EXAMPLE " 'F-TROOP' WITH DEMENTIA"

In an initial meeting with a congregation's council, there was a drab, dull, semi-hostile affective quality throughout the group. The council's leader did most of the speaking, with one or two "yes-people" elaborating. The stated problem was the order's growing numbers of elderly members and the proper disposition of their needs. The remaining members were silent, looking intently at the consultant. The group as a whole seemed both uncomfortable and defeated. One or two members expressed doubts about psychology, stating that too often it presumed to solve theological issues and that the demise of the culture had been from the self-help industry that had commercialized feel-good and promiscuous behavior.

Rather than try to provide reassurance—which actually would have been a countertransferential counterpunch to the group's displaced hostility—the consultant decided to invite the group to further develop the theme of fear of not getting help. He commented that it must be hard for the group to address such important and real problems as caring for its elderly in an age of quick-fix methods that do not always prove useful in the long run. Another member smiled and spoke about the feel-good culture and its insensitivity to problems of abortion, euthanasia, and assisted-suicide. Following this comment, there was a noticeable shifting of seats and clearing of throats. Many members appeared to be leaning forward. The consultant noticed a shift in his own body from feeling a strangled and tightened quality to one of expansion and relief. The leader then remarked how difficult it is for the order to keep new members when Americans have such little ability to make commitments, and how newcomers expect to receive more than they give. Several participants nodded and shook their heads, commenting with a rather smug certainty. Then another person who had been mostly silent added that in recruiting new members, the order had to assure them that their work as missionaries would still go on, and that they would not end up as newcomers taking care of older ones. At this point there was uncomfortable laughter among the group.

The consultant recognized that the "juice" of the theme of fear of exposing one's own vulnerability was beginning to cook in the safer "container" of the consultation process (Klein, 1992). Rather than ask the group what all the laughter might be about—a therapeutic tactic that probably would have been more appropriate in a group psychotherapy setting, though maybe not in the first session—the consultant decided to simply stick with "siding with the defense" by amplifying it. He decided to "join" the group as it commiserated, adding "Wow . . . gosh . . . and who will be looking after you guys after you've had your hands full. . . . what with the missions and now with the elderly, too. It doesn't seem quite fair, does it?" The group reacted with increased laughter. "Yeah . . . who the hell would take care of a bunch of dirty old men like us?" For a moment the mood in the group was hilarious. Several witty, self-derogatory comments continued the "don't get no respect" humor. The consultant felt amused and relieved, but also uneasy that he was being induced to see the group as one of losers.

Finally, the group played out its manic defense when one person described the council as " 'F Troop' in the cemetery business," referring to a comic TV program about a well-meaning but inept cavalry unit. The leader commented that he never realized how much effort goes into the care for the elderly until he had to "fill in" for one of brothers who regularly assists members in the wing where those with dementias are housed. The consultant had been tracking the conflict as one in which there was a manifest reactive motive of fear of being deserted or left to die. He asked the leader directly, "What do you feel your founder would have done about this problem?" Initially, several members shared that they thought the founder would have felt let down by the group's whiney style. Others then commented that maybe this was not the case, that he would have been more tolerant of their needs, just as he was tolerant of those he served in the missions.

Someone else stated firmly that the founder expected them to "die with their boots on." The consultant casually wondered if the founder would have tolerated the group's inevitable "loss of function." Would he have been let down by those who could no longer "keep marching"? The consultant noted a quiet hush in the group. There were a few members choking back more intense emotion. At this point, the leader quietly stated, "So that's what all this is about . . . we think we have to make him proud . . . we're afraid . . . we're F Troop . . . with dementia . . . in our founder's eyes we imagine this . . . it's OK to die . . . but it's 'desertion of duty' to be rendered incapacitated. We put [project] that onto our sick brothers."

The developing resolution of this very rich conflict revealed a shift in how the group expressed its reactive motive. The group began to more directly express its fear of being deserted or left for dead. The earlier comments about euthanasia and abortion now seemed to unconsciously reveal a fear, but in a disguised way. The later humorous introjective comments about "F Troop in the cemetery" were affectively charged in a way that greatly contrasted with the initial helpless anger, in which the group doubted that the consultant's pop psychology could be useful. The group now shifted its solution to a more adaptive one by discussing themselves rather than Americans who cannot commit, newcomers to the order who are ambivalent, or the consultant with his permissive pop psychology. The mirroring of the feeling within the group of being burdened seemed to reduce the need to displace the conflict elsewhere, to let the leader do all the talking, or to go on suppressing its energies in tight-faced, haughty, "pastoral" concerns about American culture. For a time the group's reactive shame could be made more acceptable in self-derogating humor.

But what was the disturbing wish that fueled this problem? The group wanted to be able to express its dependency in direct ways. The care of the elderly in the order exposed their own vulnerabilities. The organizational culture had been developed in the founder's tradition of selfless service, missionary zeal, and heroic disregard for one's self. As the realities of the membership's aging had become more undeniable, these ideals needed to be addressed in a contemporary context. The question about what the group imagined the founder would do allowed the defensive reaction against dependency within the membership to be momentarily suspended and for the group to juxtapose its founder's ideal with the contemporary demands of its context.

The balance of forces Whitaker and Lieberman (1964) referred to as the "focal conflict" can be identified as follows: The disturbing wish (here an unconscious fantasy) was to be seen (mirrored) by a proud founder; the reactive fear was that of disapproval and desertion for failing to meet his ideals; and the constricting solution was to try harder by holding themselves to an antiquated, unrealistic standard of performance. Unconsciously, aging itself meant failing the founder. The group here was able to shift in tiny ways from a solution that involved projecting its failure onto the culture (and the consultant), to temporarily doing the reverse and seeing itself as F Troop, to finally being able to consciously recognize and tolerate its mirror longings. The new solution at least temporarily allowed them as a group to let go of the restrictively compulsive and at times somewhat paranoid projections that had become the principal means of handling the conflict. Recognizing the unconscious ties to the founder's ideals in personal and contemporary terms seemed to give them a bit of fresh air.

This vignette also illustrates an essential point about organizational behavior in industry, health care, and religious communities alike: The longing in a group to make the parent proud, to enjoy the gleam in the mother's eye (Kohut, 1977) exerts enormous power that can be easily overlooked by management. For a variety of reasons, executive level decision making can essentially defend itself from recognizing its role in the motivational system, subgroupings, and basic assumptions. Likewise, it can appear to the consultant as a conflict about envy and jealousy, or about the leader's witholding from the group, and go unrecognized as a fear of vulnerability. In this vignette, the group felt too vulnerable to recognize its own pristine and untransformed wish to please the founder. In this setting, the consultation appeared to assist the community in understanding how these desires to please and to admire were connected to the founder's ideals, and how clearly or unclearly the leadership understood its members' efforts to manifest them. The group's recognition of its ties to the founder's ideals, the current assumptions about its relationship to the founder, and its unconscious fantasy about the founder helped the members to allow their own highly sensitive, childlike longings to become more accessible and more easily tolerated in conscious awareness. In this way, the founder's ideals could be upheld and restored within a more adaptive and flexible affective climate.

## SUMMARY AND IMPLICATIONS

There is a common tension in most organizations—business, politics, health care, and religious life—about how to manifest the founder's ideals. It remains like a message in a sealed time capsule, along with the group's history, frequently buried under mountains of conflict. Though the written rules and constitutions of a religious order probably make the founder's legacy more easily identifiable, organizational consultants to industry and health care may discern counterparts in corporate-level structures to the impact of the founder's legacy in the current group life.

The founder's legacy is seemingly passed on through succesive generations in two main pathways. The first is that of the leadership itself. Nodal points exist in the group's history where major decisions about new policy were implemented and where unprecedented decision making and modifications of the original group structure have required the reinterpretation of the founder's purpose and vision. The leadership of a particular era of the group's history will have some influence relative to the times, but always in the context of the founder and his or her ideals. Successive generations of leadership then grapple with how to make the founder's ideals possible in contemporary society.

The second path is through the group membership itself. The member, in his or her efforts to live out the founder's vision, bridges the past and present through fantasy and projection even where there are clear mission statements, expressions of the founder's vision. The founder's vision in turn is passed down by the group in attitudes, customs, and enculturated ways of expressing one's self and of dealing with conflict. Often the new member has no idea of what the founder was like personally, nor about what he or she sought in followers. The absence of personal experience with the founder becomes a fertile ground for even more projections about the founder from each member, often based upon the person's unresolved conflicts and developmental arrests. Each member has some theory of how the founder would have regarded him or her. These fantasy systems about the founder's ideals can coagulate into common group themes, norms, and behavior, and help feed current group assumptions (Bion, 1961).

The ways in which the members will maintain these assumptions in relation to the current leadership is contextualized by their ongoing history with one another and with successive generations of leaders, by the impact of successive generations of group events upon present attitudes, and by past as well as present group assumptions. Present fantasy systems about the leadership often contain uncanny repetitions of past perceptions and behavior of the group in relationship to its leadership and ultimately to the founder. These group perceptions are often handed down in folklore, and in chronic group assumptions. They form complementarities of membership behavior to leadership style, reflecting a reciprocity between the group's assumptions and current leadership behavior (Kets de Vries & Miller, 1991).

Yet once again, these group perceptions have not evolved in a vacuum, but in the context of the interpretation and identification with the founder's ideals and legacy as they impact upon the present group. Both leadership and membership share in ongoing reinterpretation of the founder's legacy. From an organizational consultation perspective, properly timed interventions that support the founder's legacy provide support for the group to recognize and to work with these forces. The dynamics of the founder's legacy within a religious community provide an excellent opportunity to extrapolate further principles applicable to all organizations, corporate, health care, and ministerial alike. The impact of the founder's legacy is often understated, unrecognized, and continuously present. Future studies may address these issues of the founder's legacy in organizational behavior and consultation. These currents seem to affect a wide range of the group functions in terms of conscious attitude, behavior, unconscious fantasies about leadership, and current group assumptions. The founder's legacy provides a window into the group's history of types of conflict, leadership problems, task maintenance, and affect regulation. It expands the consultant's view of the organization, allowing him or her to appreciate a greater depth of purpose across human behavior in a group setting, from administrative levels to the common worker.

## REFERENCES

Alderfer, C. P., & Klein, E. B. (1978). Affect, leadership and organizational boundaries. *Journal of Personality and Social Systems, 1,* 19–35.

Bennis, W. G., & Shepherd, H. A. (1956). *Human relations, 9,* 413–437.

Bion, W. (1961). *Experiences in groups.* London: Tavistock.

Caplan, G., & Caplan, R. B. (1993). *Mental health consultation and collaboration.* San Francisco: Jossey-Bass.

Ezriel, H. (1952). Notes on psychoanalytic group therapy: II. Interpretation and research. *Psychiatry, 15,* 119–126.

Gilkey, R. (1991). The psychodynamics of upheaval: Intervening in merger and acquisition transitions. In M. Kets de Vries. (Ed.), *Organizations on the couch: Clinical perspectives on organizational behavior* (pp. 331–360). San Francisco: Jossey-Bass.

Hirschorn, L., & Young, D. R. (1991). Dealing with the anxiety of working: Social defenses as coping strategy. In M. Kets de Vries (Ed.), *Organizations on the couch-clinical perspectives on organizational behavior* (pp. 215–240). San Francisco: Jossey-Bass.

Kernberg, O. (1985). Regression in organizational leadership. In M. Kets de Vries (Ed.), *The irrational executive: Psychoanalytic explorations in management* (pp. 38–66). New York: International Universities Press.

Kets de Vries, M., & Miller, D. (1991). Leadership styles and organizational cultures: The shaping of neurotic organizations. In M. Kets de Vries (Ed.), *Organizations on the couch: Clinical perspectives on organizational behavior* (pp. 243–263). San Francisco: Jossey-Bass.

Klein, E. B. (1992). Contributions from social systems theory. In R. H. Klein, H. S. Bernard, & D. L. Singer (Eds.), *Handbook of contemporary group psychotherapy: Contributions from object relations, self psychology and social systems theories* (pp. 87–123). Madison, CT: International Universities Press.

Klein, E. B., & Gould, L. J. (1973). Boundary issues and organizational dynamics: A case study. *Social Psychiatry, 8,* 204–211.

Kohut, H. (1977). *The restoration of the self.* New York: International Universities Press.

Levinson, H. (1991). Diagnosing organizations systematically. In M. Kets de Vries (Ed.), *Organizations on the couch: Clinical perspectives on organizational behavior* (pp. 45–68). San Francisco: Jossey-Bass.

Menzies, I. E. P. (1984). A case-study in the functioning of social systems as a defense against anxiety: A report on a study of the nursing service of a general hospital. In M. Kets de Vries (Ed.), *The irrational executive: Psychoanalytic studies in management* (pp. 392–435). New York: International Universities Press.

Rippinger, J. (1990). *The Benedictine Order in the United States: An interpretative history.* Collegeville, MN: Liturgical Press.

Schwartz, H. S. (1991). Organizational decay and loss of reality: Life at NASA. In M. Kets de Vries (Ed.), *Organizations on the couch: Clinical perspectives on organizational behavior* (pp. 286–308). San Francisco: Jossey-Bass.

Smith, A. H. (1997). Mental health consultation with religious orders. *Review for Religious, 56*(4), 387–401.

Whitaker, D., & Lieberman, M. (1964). *Psychotherapy through the group process.* New York: Atherton.

Zaleznik, A. (1984). Charismatic and consensus leaders: A psychological comparison. In M. Kets de Vries (Ed.), *The irrational executive: Psychoanalytic explorations in management* (pp. 112–131). New York: International Universities Press.

# 8

# In the Presence of the Other:

## Developing Working Relations for Organizational Learning

*Susan Long, Ph.D., B.A. Hons., M. Ed., John Newton, M.A., B. Bus., and James Dalgleish, Grad. Dup. Org. Bch.*

This chapter will describe a participatory action research project involving the state branch of a large government direct service organization and researchers at Swinburne University, Melbourne, Australia. The focus will be on those conditions that facilitated collaboration between researchers and client organization, in order that the organization might develop and learn. This has relevance also to the collaborative relations that might develop in an organizational consultancy, although a major distinction between consultancy and collaborative action research is that the latter is directed primarily to the research task of *discovery* rather than a primary task of aiding the organization to pursue its espoused aims.

Discovery and learning require a different state of mind than does the primary pursuit of efficiency and productivity. The university-based researcher has the freedom of not being directly in the employ of the organization and hence not directly subject to the authority, power, and social dynamics entailed. This is not to say that a consultant cannot achieve the independence necessary to help the organization think in new ways about its structure or processes. It depends much on the style of consultancy, the emphasis placed on learning and discovery, and the capacity of the consultant to put his or her observations forward without fear or favor. An essential aspect of research in consultancy would be the capacity of the consultant to enter the project *not knowing* about the outcomes of the consultancy. This requires an open mind and a tolerance of uncertainty. In the current project we have had to keep in mind these distinctions and to work continuously at defining the roles of the university research team members, lest we be seen as consultants giving out advice. Such a position would be countercollaborative. What it is to be a collaborative research partner seems to us more a process than a definitive position and depends upon *mature working relations,* which we define in terms of (1) clearly negotiated tasks and roles, (2) the capacity to learn through experience and action, and (3) the capacity to create and develop a facilitating and emotionally containing environment for the work. These conditions of maturity are not starting points, but are hard won through the engagement involved in collaborative work.

## PSYCHODYNAMICS OF ROLE, TASK, STATE OF MIND, ACTION, AND LEARNING

Before describing the project, we want to discuss the issues surrounding mature working relations as defined earlier, especially the development of the collaborative environment within which we believe organizational research best occurs—that is, research that will lead to creative change. Such an environment includes times and places where open discussion can occur, where the emotional life of the organization is valued, and where the many conscious and unconscious dynamic organizational processes may surface and be available for mature reflection. This really means an environment where immature and irrational impulses may be recognized and understood, rather than simply acted out or acted upon. We also believe that the development of such an environment may be facilitated by the organizational researcher, a consultant, or an organizational leader. We begin by describing our ideas about mature working relations.

## Task and Role

The first condition of mature working relations concerns clarity of tasks and roles. That the life and dynamics of the group are fundamentally structured by the tasks of the group and the role relations of the members is a central assumption for our work. Moreover, such dynamics are greatly affected by how this entity (of group or organization) exists in the mind of each member. This is because each member enacts his or her role in light of personal ideas about the whole, even though this is incomplete, fragmentary, and built up through a particular and individual set of experiences. That is, the dynamics of the group or organization are reciprocally interrelated to the way people perceive their tasks and take up their roles.

To take this account even further, we propose that the group or organization exists not simply in mental models, cognitive frameworks, or other complicated patterns of thought, but also in feelings and volitions (Long & Newton, 1997). Armstrong (1995) elaborates the meaning of "organisation-in-the-mind' to include "the emotional reality of the organisation, which is registered in him [the client] and is informing his relatedness to the organisation, consciously and unconsciously" (p. 4). In order to develop this level of interpretation and to avoid the more cognitive connotations of the term *organization-in-the-mind* we have coined the term *organization-in-experience*. The dynamics of such organizational experience are focused and refined by attention to role performance (Reed, 1976).

The idea of *the-organization-as-a-whole* is also important. Bion's work (1961) examined the way in which the *group-as-a-whole* engenders a collective state of mind in its participants. This state of mind (Long, 1996), say, a state of dependency, is internalized by the members and constitutes one form of the organization-in-experience for them. A different collective state of mind will engender other forms. Thus, one may have a role or roles within a collective state of dependency at one time, and a role or roles within a state of belligerence at another. This became evident within the current project when new tasks were faced by the various organizational role holders. The project often required organization members to examine current dynamics, including what each of the organizational subgroupings meant for the organization-as-a-whole. Working collectively at this new task led to the expression of some of the very painful aspects of the organization's work. It also led to states of confusion, anger, dismay, and sometimes hope. Members were faced with understanding their roles in light of such collective states.

## Action and Learning from Experience

The second condition of mature working relations we wish to discuss involves learning from experience and action. Action methods in learning and research emphasize the engagement of social action and its effects (Elden & Chisholm, 1994; Revens, 1983; Winter, 1989). The premise employed here is that through action the actors are effecting (that is, bringing into continuous being) the system of which they are a part. In a group or wider organization, members create their roles through the reciprocal processes described earlier. In short, the organization-as-a-whole (through its management, but also through less formal means of taking up authority and exercising power or control) may prescribe certain tasks and roles. The individual or work group may take up these roles, develop them, and create new roles according to the various organizations-in-experience available to them at different times and in relation to different tasks. Nonetheless, it is through acting and observing the results of action, short- and long-term, that people are able continuously to develop and modify their roles and to learn from them.

What is learned is not always immediately obvious. In the current project one of the work teams learned a lot about the actual implementation of "case management" by some individuals not previously identified as "good" case managers. Counter to the organization culture, this team found a way of identifying top performers by means other than the verbal endorsement of the managerial hierarchy. Despite protests from different people within the organization, they discovered what it was these people did to get outcomes. Throughout this process, the team took many risks, continuing their project despite the institutional "traditional wisdom" that wanted to qualify every discovery they made. For example, it was a risk simply to name top performers because of the (real and imagined) dynamics of envy that operated throughout the organization. They had no idea what they would learn from the experience, and many of their outcomes were unexpected. Many of the effective work practices they discovered were not part of traditional training.

In action research, the design is systematically to engage and learn from an iterative, cyclical process. Actions are planned and implemented, specific effects are observed, and the whole process is subject to reflection and discussion before further action is planned. The actions may be widespread or focused; the effects may be readily observable, long-term, or covert. In the current project we wished to explore something of the organization-in-experience for the participants and how this affected the way they carry out their roles and bring about organizational change. This exploration has led us to include the formulation of working hypotheses during the reflection aspects of the action research and action learning cycles. These working hypotheses are tested out through a further cycle of planning, action, and observation. It has also led us to emphasize the observation of experience as well as more overt observations of behavior and organization change. This in itself has led us to develop and employ methods of accessing that experience.

Finally, it led us to include attempts to think about the irrational and unconscious aspects of experience, planning, change, and action. In the field these aspects are discovered in their effects, so we were often led to explore actions and group dynamics in retrospect, or in the effects that they seemed to have on the research team or the steering group. Including an experiential working laboratory in the project design was an attempt to work with some of the irrational and unconscious aspects of experience present in the organization.

Participative action research engages a capacity in the individual to find multiple positions within the self and to take up multiple roles within the work system. Ideally this enables people to gain multiple perspectives on the task and to better understand the organization-as-a-whole because of their access to multiple organizations-in-experience. We hypothesize that when this is successfully done, a state of mind is engendered wherein collaboration and social development are possible, as is an increased capacity to tolerate anxiety, ambiguity, and other forms of social pain. Such a state of mind might include trust, a capacity to work with different power relations, and a capacity to work from a position of not knowing: a position of colabor. We recognize that like all collective states of mind, this is not achieved for all time, and that it is vulnerable to change. However, we see it as an ideal state to be achieved when possible, in a "good enough" way for work to be ongoing and generally productive. Many organizational interventions may be doomed to failure because those involved believe that collaboration is achieved through consciously formed initial agreements, when in fact it is a state of mind that is the *outcome* of work where individual and social capacities are mutually developed. We take the position that successful collaborative work is an outcome of mature working relations between individuals and groups.

We will discuss the third condition of mature working relations, that is, the idea of a holding environment, more specifically in the next section when we discuss the place of the steering committee and its work.

## THE PROJECT

Having worked as a consultant with the organization—the state branch of a government rehabilitation service employing helping professionals—one of the authors (SL) suggested a collaborative action research project focused on and aimed at (1) improving quality (of task, role, and organization) at state, regional, and local levels, and (2) developing managerial and internal consultancy roles in a framework of maximal organizational learning. This proposal was agreed to by the manager for policy and programs who became the direct organization partner in the research. Approval and agreement to be involved was given by the state manager, and both institutional parties to the agreement (the university and the organization) sought funding from the Australian Research Council, under its industry collaborative grants scheme. Under this scheme, the government matches dollar for dollar provided by the industry partner. Funding was approved in October 1995, for 2 years.

A steering committee comprising three members of the university research team and six organization members was established to plan, guide, oversee, and evaluate the project in an ongoing way. This committee met monthly or more frequently when required.

The research occurred over five stages:

1. Initial (voluntary) interviews, focused on role, task, and quality, linking these with the perceived primary task of the organization, were conducted by the university research team with a sample of organization members across all levels of the organization (76 interviewees).
2. Analysis of the interview data and feedback of major themes to the steering committee and to all participants occurred.
3. A residential "participatory action research working laboratory" took place over 3 days. It included an experiential component, further exploration of interview findings, and the development of action learning projects by teams of organization members.
4. Teams worked over a 12–15 month period on different action learning projects.
5. An ongoing working model of quality improvement comprised the final stage.

## Providing a Holding Environment for the Project

We have noted that researchers of organizational dynamics or consultants frequently engage in collaborative efforts with the members of an organization to bring about organizational learning or change and to understand the antecedents and effects of change; yet the very process of collaboration is itself often taken for granted, especially when it occurs successfully. Alternately, when the collaboration is under stress or breaking down, the processes considered are often framed in terms of resistance, whether this be to the "outsiders," to change, to management, to the organizational culture, or to its environment. Might there be other ways of framing such changes in the collaborative relationship? Some of the work done on understanding the breakdowns in collaborations, between researchers and organizations or between consultants and organizations, has been around the difficulties of establishing initial working relations. Other work has been done examining the politics of ongoing relatedness (Casemore et al., 1994; Mirvis & Berg, 1977). However, we need also to study the process involved in successful collaboration.

In designing the project, we understood the need for a setting that could allow and contain negative as well as positive aspects of the collaboration so that these could be understood, learned from, and worked with alongside the primary work of the project. Drawing on models and ideas from psychoanalysis and therapy, where collaboration between therapist and patient is required in order to understand and work with the patient's problems, others have talked about such a setting as a "facilitating" or "holding" environment (Bridger, 1990; Miller, 1995; Stapley, 1996; Willshire, 1997). What is meant by this is a setting that itself can be trusted to provide physical and psychological safety to those willing to experience and explore issues that might normally be anxiety-provoking, politically subversive, or countercultural within the organization. Such a setting allows exploration of ideas and feelings that emerge during work yet which in less exploratory or safe settings become suppressed or operate in an underground manner. A good holding environment allows for a containment of such experiences and explorations, so that they can be integrated for work rather than for counterwork purposes. For example, anger with a manager might be talked through with him or her rather than left to smoulder in resentment (and possible unconscious or even conscious sabotage to the work expressed through lateness or absenteeism) because the issues involved cannot be broached. In fact, all cooperative and collaborative work requires a form of holding environment in order to facilitate the human interactions involved and the learning from their consequences. Where the purpose is collaboratively to study and understand organizational dynamics, as in the current project, the holding environment needs to include a facilitation of all aspects of the collaboration for these purposes. Further, rather than requiring a holding environment simply for what already exists, a new collaboration requires an environment that will foster that which currently does not. This was a challenge for our project design.

The steering committee was set up to provide a venue where the collaborating partners could begin work together. It was hoped that the partnership between the university researchers and the organization could be extended and worked with in many different ways as the project progressed; however, there needed to be a central group that could take on the tasks of designing, steering, monitoring, and evaluating aspects of the research in an ongoing way and in light of the progress occurring within the project. It was to be a forum for discussion of the activities within the project and a place where there was a sense of the whole of the project.

Both institutional partners had their own needs to be met by the research. The organization wanted to improve the quality of work done by its professionals, but they also were going through many changes to their structure, the way they were funded, and the types of clients they had, and this had an impact on the way that management took up the opportunity of the research. For example, they were preparing to become either a statutory authority or a government-owned enterprise—either of which meant they had to have increasing independence from government, would be operating under a funder/provider division, and would be located in a competitive environment where once they had a virtual monopoly. Moreover, the organization members, including management, are dedicated to the philosophy of minimizing job loss through the change process, counter to the New Zealand experience of change, which often involved dramatic changes in organization structure and personnel (Spicer, Emanuel, & Powell, 1996). Such radical change processes deal with "resistance" by shedding the so-called resistant people. This organization wanted to achieve change through other means. Despite a climate of financial cuts, it was agreed early on to call the project "the organization growth project." This seemed to symbolize the hope that growth could occur in ways other than size or funding.

On the other side of the collaboration, the university wanted the chance to engage in research. Having recently moved to the status of university from a college of advanced education, building a research culture and gaining income and kudos from research was important. Beyond this, the research team had its own particular way of working and its own particular way of conceptualizing organizations, as described earlier in this chapter. Success for the team would mean being able to work with a model of learning from experience and from the attempted exploration of unconscious processes operating while organization members went about their tasks. We strongly believe that the interests and needs of both parties to a collaborative venture—research or consulting—have an effect on the project undertaken. The researcher or consultant must look to himself or herself in understanding the organizational change under consideration.

As the major containing environment for the project, the steering committee needed to have strong authorization for carrying out the work of the project. The university team had expert knowledge and a neutral or objective perspective. This was valued, particularly as some successful consultancy work had been carried out with the organization by members of the research team. The organization members of the committee had inside knowledge of the organization, and the authority to make detailed decisions about what might be done within the project. Moreover, many members of the senior decision-making body within the organization were on the steering committee. For a successful collaboration to develop, both parties should feel that their needs are being worked toward.

The initial relationship for the research was between one of the authors (SL) and the manager of policy and planning of the organization. It seemed important to have the state manager strongly involved, not only because he held authority for the branch, but because his position allowed him to have a close understanding of the national situation for the organization and the changes that were occurring at that level. He agreed to chair the steering committee, and he responded to the research proposal. Although he made it clear that the way of doing things outlined in the proposal was not his usual way, his preparedness to "give it a go" arose from his recognition that the organization needed to engage new approaches to issues of quality. He believed that the old established ways would not get them much further, given the sweeping organizational changes upon which they were about to embark.

It was the manager for policy and planning who had the better understanding of the concepts, models, and methods used by the university team, being a graduate of the program in organization behavior conducted at the university. As the project was developed, and in its early stages, he played an interpretative role, translating between the two collaborating groups. As he said later, "the public service way of doing things is very hierarchical, and this new way of working required a significant amount of explaining in my colleagues who with one exception could not see the benefits straight off." During a large part of the first year of the project this manager was ill and the major working relationship between the partner institutions was maintained within the steering group broadly, and between the senior researcher and the state manager more specifically. But by this stage the project was well on its way, and the two groups had to build their own bridges and make their own translations.

Despite the state manager's ambivalence about the project he was very supportive at many levels, including protecting it at the national level where other projects on quality may have put an end to this one; supporting the project financially; taking up the role of chair of the steering committee; attending most meetings, where he took an active role, officially reinforcing the importance of the project to members of the organization through electronic and other communications; and devoting his own time to the project in other ways when necessary. We have come to see some of his stated, and at times acted, ambivalence as representative of public service conservatism, and he has clearly stated that he sees himself as a career public servant, as do many of the senior managers. However, as the project has progressed, other members of the steering committee, including members of the university research team, have also experienced ambivalence toward the project and their part in it. This has been expressed through, for example, a female case manager's withdrawal from the steering committee, and through the experienced marginality of one member of the university research team. It may be that the steering committee dealt with collective ambivalence and associated anxieties through the experience of just a few of its members, leaving others to take a strong positive stance. Such a "splitting" defense was also present in the wider organization with respect to ambivalence about the broad organizational changes mentioned earlier. As one manager put it, "Half the organization has its foot on the brake while the other half has its foot on the accelerator." In order to get on with the research task, the project had to provide the space to explore this dynamic, expressed as a parallel process in the organization, the steering group, and the research team.

The members of the university research team had to attend closely to their own experience in order to learn how this dynamic operated with them. The following example may serve to illustrate how the design and process work on the project enabled such an exploration.

## Example

We have said that maturity in work relations depends on a capacity clearly to negotiate task and role. This is an ongoing process, not an initial point nor an outcome. Hirschhorn (1988) describes a retreat from the role boundary (with a task or with another role) as a defense against the anxiety generated by the task. For example, an employee may draw back from reporting or discussing something that he or she regards as a mistake made by an organization "superior." The anxiety aroused by crossing someone in authority (and, of course, the perceived or imagined repercussions) leads the person to withdraw from interaction or perhaps from carrying out a rectifying action. The outcome may be loss of quality or production or initiative for doing something perhaps new and creative within the work. The withdrawal is protective of the individual, at least in the immediate sense of "pulling one's head in." We believe that such a retreat from the role boundary means that the ongoing task is not renegotiated, nor is there a chance to renegotiate and change the nature of the roles involved. In this case the superior has not been given the chance to learn from a mistake. Moreover, the fantasies surrounding authority have not been tested in reality and are likely to continue unabated. This provides a situation likely to increasingly support a culture of not challenging authority. If, on the other hand, the employee does take up the full authority of his or her role, questions the superior, and receives a rebuke, then the roles have been renegotiated with a little more certainty, perhaps to the detriment of the work. That is the risk. However, given ongoing circumstances—one being the reality that continual mistakes or being noncompetitive will cost the company or organization in one way or another—risking renegotiation of task and role brings the people involved closer to the realities of their work and closer to a more satisfying working relationship. Such dynamics and choices are present in the organization studied, and in the university.

An early indication of the need for continual role negotiation occurred in the university research team several weeks after the project had been "launched" in the wider organization and when the steering committee was beginning to make decisions about the interviewing/data-gathering phase of the research. The two members of the team who were full time academics (SL and JN) became concerned about the contribution of the third member (JD). Their concern emanated from his relative silence in discussions and his lateness or absence at short notice from scheduled meetings. The senior researcher (SL) broached the matter with him, and through discussion it emerged that he felt treated as less than equal with the two university-based members. As evidence he stated that in introductions to members of the collaborating organization his status as a sessional lecturer at the university had not been mentioned and during the project launch he had been reduced to the role of chauffeur when visits had been made to various regional offices. Subsequently, he believed that he was being treated as a junior rather than as an equal member of the team.

This issue brought up for mutual consideration the multiple roles that could be manifest or latent within the team and that were contributing to the differing organizations-in-experience. For instance, the researcher (JD) who was relegated to the "junior" status was concurrently enrolled as a postgraduate student at the university and was being supervised on a separate research project by his sometime colleague (JN). JD's personal research project was proving to be a challenging and painful experience, yet it was his initiative and risk taking as a "student" that had stimulated the invitation for him to join the collaborative research team. He had been given the opportunity because he was a student of the university who was seeking to learn more about psychodynamically oriented action research and because he had the capacity to undertake particular tasks within the collaborative project. The tensions between the roles of teacher, student, and research colleague could not be defined away but had to be brought into the relatedness of roles and task, most particularly in this case, because the university team was seeking to develop a relationship of coresearchers with the members of the organization. It may be that the desire of JD to be properly recognized as a sessional lecturer at the university was not just about his status within the research team. It reflected, too, the very positive transference from organization members onto the university representatives. At the launch of the project many organization members expressed a sense of pride and importance that they were going to be involved with a university on a government-funded research project into the future of their organization, and there were expressions of relief that this was not to be another short-term consultancy with foreseen conclusions. It was to be conducted by independent, outside researchers.

The power of this transference was reflected by the feelings of JN who assumed, without discussion, that only he and SL could fully represent the university and who, by his actions, left little space for JD to take up the fuller role previously assigned to him in the research. This relationship was complicated further by the fact that SL and JN were seen by some of their colleagues as a strong "pair" in their university work and JD was struggling to find his role as a colleague in the face of this.

Exploration of this dynamic allowed the research team to begin to understand and explore some parallel issues in the organization. In fact, we hypothesized that issues within the research had activated and brought to the fore some of our own dynamics. The negotiation of multiple roles and tasks was problematic for the organization. A striking example is the role of Senior Adviser, a senior professional role, with professional responsibilities and accountabilities, and with the additional role of internal consultant to all line levels within the organization. The complexity of the role meant that there are often conflicts between the differing role aspects, such as when a senior advisor might have to do file audits with the direct professional staff (seen as a policing role within the organization) while also acting as a consultant to the work of a case manager, or to the case manager's manager at the unit level. People working in this role constantly have to negotiate the tasks that they take up, the authorities they hold, and the role relations that they have with others. Also, the role of Administrative Officer in the organization, although seen as invaluable, leaves its incumbents feeling as if they are "juniors" within the organization. Issues of hierarchy, relative power, and exclusion, then, were reflected in both the organization and the research team.

## The Culture of the Steering Group as a Reflection of the Culture of the Organization

We have described how a holding environment allowed the exploration of research team dynamics as a tool for further understanding the organization researched. It was also important to explore the dynamics of the steering group. As with the case of the research team, this had a dual purpose. First, it threw light on the organization culture more broadly. The steering group acted as a kind of organizational microcosm. Second, it was through the collaborative and exploratory work of steering group members that more mature working relations were developed for the project to proceed in other parts of the organization, for example, in the working laboratory and the action learning projects.

While having a steering committee seemed to fit easily into the culture of the organization, insofar as most of their usual projects had what they called a reference group, such a fit was not always helpful. Reference groups had a particular way of operating, making the introduction of new ways of working in the steering committee difficult. For example, a reference group, despite having the ostensible task of guiding and evaluating a project, has the implicit task of representing the various interest groupings or structural configurations within the organization. This is associated with the task of seeming to steer a fair course between different interests. A reference group might be criticized for being politically biased, often by leaving particular groupings poorly represented. Such criticism, directed at the straw argument of representation, often covered for other issues that were not directly broached. In the establishment of the steering committee, there was a strong push to have the different role groupings within the organization represented, so that the project might be politically acceptable. Although this might be an important end in itself, it tended to overshadow the need to have people on the steering group suited to its particular role in this project.

The stated feelings underpinning these reasons for selecting reference group members tended to be those of not wanting to hurt the feelings of others, nor wanting others to feel that they are being treated "differently." The ideology is one of fairness for all, a kind of forced egalitarianism that may well be a defense against the experience of envy (Schoeck, 1966). In the organization, this ideology slips at times, finding other-than-direct ways of approaching clear differences, such as of experience or expertise among the staff. This then becomes fertile territory for underground political intrigue.

Such a dynamic may be connected to the task of working with disabled clients and the mandate of the organization to follow government policy in integrating the disabled into the wider work culture. Although following a worthy ideal, it is much more difficult to get jobs for people with some disabilities than for others. For example, what is nowadays termed "psychiatric disability" might cover a person with a chronic psychiatric problem who has rarely if ever been employed. In the context of the organization becoming more "businesslike," one person said, "How do you make money out of the mentally ill?" There is an intrinsic contradiction within such a task, because aspects of it seem quite impossible. It is as if the task idea, that is, the task that organization members have in mind, has become a corrupted task (Chapman, 1996) in the case of psychiatric clients. That is, the task of helping them to gain work has become the task of using them for organizational gain and political expedience. Government directives have moved the organization away from a prior, albeit minor, aim of helping people gain "independent living skills," to a greater emphasis on vocational outcomes. In the long term, such clients may not remain with the organization. In the interim, staff specifically trained to work with these and similar people, who entered the organization with social justice values, find themselves caught in a bind.

This is reflected throughout the organization. Yet although the ideology of "everything and everyone equal" is recognized as somewhat hollow, it is still visibly played out. The reality of authority and power behind the ideology is different. For example, there are power cliques within the organization based on geographical regions and key roles within these areas.

Whether or not the steering committee for this project should be a representative body was discussed. Despite attempting to think about what was right for this project, the established organization-in-experience became reestablished in the steering committee. Membership was largely representative of roles as interest groupings, rather than roles as pertinent to the task at hand. Also, as often occurred with reference groups, it was some months into the project before a case manager (the direct service role) was invited onto the committee.

This dynamic of unconsciously reestablishing the usual organization-in-experience seemed to spill into the project activities. The number of interviews to be conducted as part of Phase 1 continued to grow as steering group members felt that all roles should be interviewed in large numbers, and pressure was put on the interview team to enlarge the interview sample. There seemed to be compelling forces drawing the university team into the culture of the organization, rather than into a partnership.

Although dealing with the pragmatics of introducing the project more broadly in the organization and planning the interview phase, the early meetings tended to be formal, with people tentatively getting to know one another across the partner institutions. Little processing of group dynamics occurred at this time. It was mainly the overall design of the project; the investment of hope, time, and energy that the organization had put in; the promise of a report from the interviews; and the future working laboratory that acted to keep the project alive and continuing.

The time for exploration of the dynamics came during Stage 2 of the project, when the university team provided feedback of the results of their interviews to the steering committee. Three half-day sessions were set aside for this process. Certainly, this phase led to a deepening of the relationship between the steering committee members. The university research team had spent a lot of time reflecting on the results of the interviews and on their own internal team dynamics. Although some of these dynamics clearly belonged to the team and reflected other working relations that they had within the organization-in-experience of the university, as described in the preceding example, other dynamics seemed to reflect issues that were important to the organization. Working through these dynamics strengthened the understanding that the team members had about what was being said in the interviews and helped them to form several working hypotheses about the organization, and about the way the dynamics involved were occurring within the steering committee. It was these working hypotheses, together with the interview data, that the research team was able to bring to the steering committee during the feedback sessions. Examples of working hypotheses follow.

*1.* Becoming "businesslike" means different things to different people in the organization; certainly, different groups focus on different aspects of being "businesslike." Many have unverified ideas about what constitutes "businesslike" behavior and work practices in the private sector, yet they try to envisage such practices. Our working hypothesis is that there are no clearly defined ways of translating the general idea of becoming "businesslike" into specific work practices. The imperative to do it is present but the models for operationalizing it are few. It is often a hit-or-miss approach. How might the realities of businesslike practice be discovered and how might these be appropriately adapted to the organization's needs, environment, and market?

*2.* Moving from a two-person focus (case manager and client) to a three-party focus (case manager, client, and customer) poses problems for case managers' learning (of this new role) from experience. Such learning requires support, mentoring, and an opportunity to discuss ongoing problems and issues as they arise. We hypothesize that the anxieties associated with such learning are exacerbated by a felt lack of guidance and support, which in turn makes discussions about values surrounding the changes fraught with anxiety. What sort of learning environment might be developed to assist in this change?

*3.* Staff are clear about the need to change and are largely unresistant to this idea. The problem lies more with the specific detail of how to change while keeping the purpose of the organization intact, and while keeping the identity and integrity of personnel intact. This reflects two issues: a technical (how to) and a values (with what effect) issue. We hypothesize that these two issues are effectively intertwined and require being worked on together at each point of change. How might personnel discover the specific details of change required for their roles? And how might dialogue around technical and values issues be promoted?

In light of the feedback, the committee was able to address not only issues surrounding the organization as a whole, but also some of its own dynamics. This was a risky process, because what was raised threatened the established ways of thinking. At times it worked well. For example, managers on the committee began to see that what they had viewed as a message "not getting through to others in the organization" was clearly not the case. The message (concerned with changes that had to occur) had reached the members of the organization; their response to the message, however, was complex. Management had read complex responses as if they meant people had not heard. Such responses included staying quiet and getting on with work as usual, or being perplexed about the implications of the message, or simply not knowing how to go about the operational changes needed.

At other times, the "feedback" process seemed to lead to a restressing of past, not particularly useful, ways of doing things, which had to be challenged. For example, the simple perception that the organization was divided into those who were for change and those who resisted (the splitting phenomenon mentioned earlier) kept being discussed and concretized into several metaphors despite the evidence that the situation was more complex. This would lead to ideas of resolution, metaphorically expressed, for instance, as "pruning out the bad," or "putting bombs" under some people who were seen to be on the "bad" side of the split. This perception was held not only by steering committee members, but also by many throughout the whole organization, as we found in the interviews. Moreover, it tended to become a self-fulfilling prophecy, with some people acting out resistance even though many individuals recognized in their own experience that the situation was far more complex.

The steering committee, through the feedback sessions, worked on their own perceptions and moved from a more simple view to an increasing acceptance of complexity. Interestingly, since the commencement of the project, the senior managers have changed the composition of the major decision-making group in the organization. It now has a far wider membership, including an Administrative Officer—a role that many saw as important but with little authority or power. It is this new body that approved the action learning projects. The steering committee also expanded, not to better represent interest groups or structural configurations, but to include people appropriate for task.

In both cases, this move from the idea of political representation to selection for task was influenced by the work within the project.

## The Working Laboratory

The working laboratory allowed the process work of the steering committee to be extended out to a wider group in the organization. It marked a further step in the development of mature working relations between the university research team and the organization. That is, many more organization members: (1) worked at negotiation of task and role within the project, that is, developed action learning projects; (2) developed their capacity to learn from experience and action; and (3) worked at further developing their own team capacities, including their processing of their work as it proceeded.

Designed to take place over 3 days, the working laboratory had a task "to explore and study the emergent dynamics of the organization in relation to the issues facing it in terms of quality work and organizational learning." The notes given out to members went on to say: "The task will be worked at in two main stages. The first will be through an examination of current and (anticipated) future role identities, and the dynamics of transition between these identities.... The second stage will involve the development of working hypotheses about the organization and its movement toward an anticipated future, and from this, the initial development of action learning projects."

The design included an opening plenary session followed by a day and a half of an experiential intergroup event (Stage 1, described later). Some reflection and plenary time was provided to process the experiential work. A second day and a half consisted of a work team event (Stage 2) designed to enable self-selected teams to begin the design of action learning projects that could be carried out within the organization over the next 15 months. The steering committee met during this event, with the task of selecting five or six of the projects that would then be recommended to the organization's decision-making body for implementation. Teams presented their projects to the steering committee during the final session of the work team event and the committee then made its decision. During this latter day and a half, time was also available for reflection and for plenary discussion. The 3 days ended with a final plenary for reflection and discussion of the working laboratory as a whole.

The first stage proved to be extremely difficult for the members to grasp, and their efforts generated a volatile emotional climate for the shift to the second stage of the working laboratory. The issue of authority within the laboratory experience was also vexed. It is significant that the state manager informed us indirectly and only shortly before the working laboratory that as a consequence of some family responsibilities he would not be participating. He had formed the idea that his absence would not be important since he assumed everyone would be out of their usual organizational roles anyway. After some strong persuasion from his colleagues on the steering committee he reconsidered his commitments and made himself available for the second stage of the working laboratory. Thus, he arrived during the lunch break between Stages 1 and 2.

In the second stage of the working laboratory, members were requested to form self-selected groups on the basis of their mutual identification with a dilemma facing the organization and their willingness to develop a proposal for an action learning project to tackle this dilemma. All members of the laboratory had previously been given the research report with the hypotheses described earlier in this chapter. It was understood that the steering committee would assess the merit of the proposals and select some of them for recommendation to the organization's business strategy committee for support and implementation. It was stressed that competition between projects for selection was viewed as a desirable means of improving their quality. The steering committee wanted to have some choice.

Two of the steering committee's tasks were to think about the extent of the competition they wished to generate between action learning groups and the process to be employed in selecting the "winning" proposals. Such explicit promotion of competition was countercultural for this organization but it was deemed fitting now, given the need to prepare the organization for its movement into a competitive marketplace. The consultant to the steering group kept pressing this point as members voiced concerns that there was not enough time to do justice to the action learning proposals and that people could be hurt and the project damaged if expectations were dashed.

The courage was found to have an open selection process led by the state manager and the principal researcher. The members responded to this by producing a high level of competitive bids for projects, which were outlined by team representatives within strict time limits. Working within the competitive environment was stimulating, and the steering committee was faced with difficult choices about which projects should be endorsed. There was a lot of enthusiasm generated. At the conscious decision-making level, some of the projects seemed to hold synergies that suggested they be brought together. This would allow each of the ideas formed to find expression. This meant at an unconscious level that the effects of competition were avoided by the merging of some projects so that everyone "got a guernsey," again reflecting the organizational dynamic of equality for all. Yet one proposal, suggested by a single person, "fell off the end" of the process by not surviving, nor really becoming amalgamated with any other. The organization-as-a-whole really did not know quite what to do with this situation. It was as if a firm rejection could not be given to anyone. However, in its absence, a firm approval did not seem to be forthcoming, either.

Nonetheless, the committee was impressed by the amount of work done by the teams and their capacity to present quite detailed working hypotheses (about the organization) and possibilities for projects within a short space of time. Learning and efficiency seemed to be occurring together. Was this despite some of the confusion? Was it because others were able to take up their authority in the face of an uncertain management which, however, also gave them a new space and a new organization-inexperience? Certainly, steering committee members found that they still had a lot to learn about working together under pressure. The capacity of members to represent the work of the group-as-a-whole was challenged throughout the event.

It was under these circumstances that a new level of judgment and trust was developed. Members had a strong basis on which to judge each other. After the working laboratory, it was not so much that steering committee members trusted each other more per se. However, when differences arose between members in the course of their organizational work, these began to be talked about and resolutions sought through open discussion. For instance, one project led by a member of the steering committee began to proceed without consultation with another member of the committee who had line responsibility for the region within which the project was to take place. This led to difficulties and hostilities in meetings where both were present. Yet it was within the steering committee that they were able to discuss and begin resolving this.

One "holding" or "containing" aspect of the laboratory, stimulated by our working hypothesis about the need for new forums for communication, turned out to be provided by the plenary reflection sessions. These sessions became partly reflective, partly working on here-and-now experience, and partly attempts to conceptualize the experience. They provided quite a powerful space where some members were able to put into words their own organization-in-experience in ways they had been unable to formulate previously. Most importantly, this work was done openly, in public. For example, one member said, with difficulty and evident courage, that she did not know how to do business and often felt that she would prefer to remain within a public service that provided her with a secure job rather than have to work within a competitive semi-autonomous body. Furthermore, this really could not be said within the organization given current rhetoric, despite her belief that many felt as she did.

We have learned from this project that the bringing of ideas, thoughts, and experience into the public arena of the organization is extremely difficult, although believed to be important. We suspect, however, that for such processes to occur successfully the state of mind associated with discovery and learning must be engendered. This would be antithetical to the state of mind associated with the covert politics that underlie the defensive "we are all equal" stance. One of the action learning projects called "the listening post" (Khaleelee & Miller, 1985) has grown out of the work of these plenaries and now provides an organizational forum where a cross-section of the organization meets bimonthly to air thoughts and make connections in public, difficult as this is.

## THE COLLABORATIVE STATE OF MIND

The questions that we wanted to address in this chapter surround the nature of collaborative working relations. Our interest here is that although the collaborative action research project has a major research task, that is, a substantive task, the social setting for the research is as crucial to the completion of the task as the methods employed in the research.

### Types of Relatedness

As the research proceeded, we came increasingly to see the processes of collaboration and mature working relations as complex. Bion (1970) outlines three types of relatedness between the genius and the establishment: the new idea and the established set of ideas available to a group. These may be relevant to relatedness between any two entities. The three types, taken by Bion from biology, are commensal, parasitic, and symbiotic. Commensal relations allow the two entities to develop alongside one another, perhaps sharing some aspect of the environment or the relationship but basically leaving each other alone. Parasitic relations involve one entity subsuming the other in both an instrumental and destructive manner. Symbiotic relations indicate the possibility for creative development of both entities. Each forms part of the environment of the other in such a way that mutual benefit can be derived. Together they form a microclimate, beneficial to both. We can think of collaboration in these terms.

1. Collaboration could involve development of the parties in a common climate where each derives benefit from that climate but essentially develops independently (commensal). In terms

of the collaborative research project, both the organization and the university might gain from the research occurring. Both might be able to meet their needs within the project, without really affecting what the other is doing.
2. Collaboration might involve the subsumption of the needs of one partner by the other. It might do this to the extent that its needs become met to the detriment of the other (parasitic relations).
3. Collaboration might involve the development of a micoclimate where the development of one party actively promotes the development of the other (symbiotic).

Alongside this typology of relatedness, we find useful the Weberian distinction between instrumental reasoning and value reasoning (Weber, 1968). A relationship may be forged for instrumental reasons or for reasons of promoting a particular value or set of values. As the initial values of the two parties in a collaborative venture differ, it seems that any collaboration of the symbiotic type will require a renegotiation of values on both sides. This is exactly what would happen were the parties to learn from each other. A merely instrumental relationship might occur within parasitic (destructive) relations, or within commensal (benign) relations. Values may shift as a result of these latter relations, but we suspect that learning would be of a type that simply reinforced past or current prejudices, in contrast to higher order learning.

## Toward an Understanding of Colabor: Working in the Presence of the Other

A hypothesis might be formed that the collaborative relations within the project were primarily commensal. The evidence would be that the authority relations illustrated not so much joint as divided authority. Sometimes authority felt split, as between responsibilities for the first and second halves of the working laboratory, where the university staff had authority in the first half, while the organization members held authority in the second half. Sometimes it seemed closer to colabor, as within the plenaries. Second, the separate interests of the research team in working with a "here-and-now" or learning-from-experience methodology, and the needs of the organization for "outcomes" seemed to run in parallel. This became closer to a symbiotic relation the more the steering committee took up its responsibility for the overall success of the working laboratory and for the project more broadly. For example, several of the steering committee members and other members of the organization worked hard to convince their colleagues that it was worth persisting with the working laboratory at those times when the experiential work became taxing or confusing. Third, both the university team and the organization members seemed to be working unconsciously on the relations between followers and leaders in their respective groups. This was evidenced in the example given earlier about the university research team, as well as at the laboratory. Once again this seems evidence of commensal relations, where both partners work side by side on their own issues, sharing an environment but not necessarily affecting one another greatly. Such a scenario is not uncommon in some consulting relationships where the consultants are primarily focused on developing or applying their own "packages" of consulting tools, in parallel to the developmental needs of their organizational clients.

Looking beyond such initial impressions at the process of collaboration we must pose to ourselves the question: What would be evidence of the two organizations learning from one another? Might it be that the project-as-a-whole provided a microclimate within which issues for both institutions of changing into customer-focused businesses may be worked on at an unconscious level? Might it be that *in the presence of* the other organization, each is able to try out new ways of thinking about what such a focus means for them? Also leadership and its effects have been important for the university team (within the team and more broadly in the university), the organization (at all levels), and the project steering committee. The questions of what is being learned, by whom, and from whom are not easily answered, especially when much of that learning is tacit and not easily accessible for reflection. Just as much of what we learn as infants in the presence of our parents is unconsciously internalized, so might much of what organizations learn in the presence of each other occur in ways that are difficult to study. The same might be said of learning in the presence of a consultant or consulting team.

We are reminded of the many times the university team members have learned about the situation facing the university from drawing parallels between the political and business environments of the government organization and their own, as, in particular, the environment of increasing independence from recurrent government monies with an associated pressure to develop full-cost recovery programs from a competitive market. This has led both the health sector and the education sector to strategically organize with a stronger customer focus than previously held, and to operationally organize in terms of cost-cutting through staff rationalization, faster "throughput," and the implementation of flatter management structures. In particular, we have noted the increasing stress and accountability placed on frontline workers who seem to carry the hope for increased productivity, yet who seemed most stretched by the changes. This parallels the increased pressure put on frontline teaching staff at the university, who are expected to work with larger groups in shorter time frames, and yet increase the quality of programs in order to compete with larger, more established universities both in Australia and overseas.

Both organizations face these pressures from a traditional value base of professional services, where the professions have had prime authority in deciding what is required for their clients or students. They now face increasing pressures from a variety of stakeholders in the provision of services, not least of which emanate from the needs of industry, itself competing under a free market philosophy. Both organizations contain ambivalence about the changes occurring within them, and contain members who believe that the primary tasks of human service or of education have been compromised quite seriously.

The opportunities for examining the effects of various responses to these pressures have been numerous. Members of the steering committee have reflected, for example, on the uses and abuses of information technology in change processes, in particular, on the difference between "information" given over the electronic mail system and its interpretation by organization members. Increasingly, both organizations are moving into a network mode where E-mail and teleconferencing may provide as much interchange as face-to-face meetings.

Yet, in line with this, it is apparent that new spaces and forums are required for the expression of organizational experience and for the renegotiation of the roles emergent from the new organizations-in-experience. This need was rather poignantly expressed by one of the regional managers, who found himself having to chair a statewide meeting. A statewide meeting employs huge organizational resources in terms of time and travel, and members expect to get strong paybacks for that in terms of learning from others the broad state of play. This manager felt that somehow he had to conduct things in a way that met a nascent need not articulated in the organization, and, yet, the formal agenda did not seem to allow for this. Such a mismatch aroused an enormous amount of anxiety and responsibility in him.

Many of the steering committee meetings have addressed this issue of communication. Whereas the organization members, like university administrations, often held a model of communication as transmission (of preexisting data), the university research team often held a model of interaction, where information develops within exchanges. Holding on to this latter view, in the face of organizations that anxiously wish for some preexisting certainties, has been difficult. However, it has been by exposure to such a view that the organization has begun to develop new ways of sharing and creating knowledge.

## CONCLUSION

Our experience in the project described in this chapter confirms the need for a suitable ongoing forum where collaboration can be continuously renegotiated and monitored. Not only is this important for establishing and maintaining mature working relations, but it also throws light on the substantive work of the project. Discovering parallel processes in the collaborative partners can aid in the further understanding of how the effects of these processes are differentially handled. Framed in this way, the collaborative partners may come to see their relationship as systemically linked to issues in the organizations that they represent. This may include the institution of "research" or "consultancy" as well as an organization. Focus on the relationship can then be regarded as a tool for understanding broader organizational issues.

We said earlier that learning may occur unconsciously *in the presence of* the other. It seems to us that collaborative learning occurs when the other is present to us in a *state of mind* compatible with our own—perhaps there is something about the nature of trust in this. Another way of putting this is that symbiotic collaboration only occurs when the partners can *be* in the *presence of the other*, that is, they can bear to listen to each other in a way that makes them available to each other for learning. Moreover, in order for this learning to be available for reflection and further conscious learning, and in order for the collaboration to be linked to the task at hand rather than being a collusion around unspeakable dynamics, the collaborative process requires ongoing scrutiny. Finding the space to explore this at both conscious and unconscious levels, we believe, leads to a deepening of the relationship and its possibilities. One of the managers described his experience of the collaboration as having quite a degree of intimacy, yet maintaining a separateness between the two partner groups. This had helped his learning and, as he put it: "I never did understand what was meant by a strategic alliance. The idea of collaboration as we have developed it seems more real."

## AKNOWLEDGMENTS

This project was made possible through a grant provided under the Industry Collaborative Grants Scheme of the Australian Research Council and the organization described in the chapter. We acknowledge the helpful comments of many others, especially Ranjit Bhagwandas, Jane Chapman, Chris Foley, Geoff McInnes, Charles Langley, George Manoussakis, Mandy McGuire, Colin

Roberts. We thank all members of the project—too many to mention, but each contributing.

# REFERENCES

Armstrong, D. (1995). *The analytic object in organisational work*. Paper presented at the Symposium of the International Society for the Psychoanalytic Study of Organisations, London.

Bion, W. R. (1961). *Experiences in groups*. London: Tavistock.

Bion, W. R. (1970). *Attention and interpretation*. London: Tavistock.

Bridger, H. (1990). Courses and working conferences as transitional learning institutions. In E. Trist & H. Murray (Eds.), *The social engagement of social science: Vol. 1. The socio-psychological perspective* (pp. 221–245). London: Free Association Books.

Casemore, R., Dyos, G., Eden, A., Kellner, K., McAuley, J., & Moss, S. (Eds.). (1994). *What makes consultancy work: Understanding the dynamics*. London: South Bank University Press.

Chapman, J. (1996). *Hatred and corruption of task*. Paper presented at the Australian Institute of Social Analysis, October Scientific Night, Melbourne.

Elden, M., & Chisholm, R. (1994). Emerging varieties of action research. *Human Relations, 46*(2), 121–142.

Hirschhorn, L. (1988). *The workplace within: Psychodynamics of organizational life*. Cambridge, MA: MIT Press.

Khaleelee, O., & Miller, E. (1985). Beyond the small group: Society as an intelligible field of study. In M. Pines (Ed.), *Bion and group psychotherapy* (pp. 354–385). London: Routledge & Kegan Paul.

Long, S. D. (1996). *Psychoanalysis, discourse and strange lists: These are a few of my favorite things*. Paper presented at the Symposium of the International Society for the Psychoanalytic Study of Organisations, New York.

Long, S. D., & Newton, J. (1997). Educating the gut: Socio-emotional aspects of the learning organisation. *Journal of Management Development, 16*(5), 284–301.

Miller, E. (1995). The healthy organization for the 1990's. In S. Long (Ed.), *International perspectives on organizations in times of turbulence* (pp. 1–13). Melbourne: Swinburne University.

Mirvis, P., & Berg, D. (Eds.). (1977). *Failures in organization development and change*. New York: Wiley.

Reed, B. (1976). Organisational role analysis. In C. Cooper (Ed.), *Developing social skills in managers: Advances in group training* (pp. 1–11). London: MacMillan.

Revens, R. (1983). *The A B C of action learning.* Bromley, Kent, U.K.: Chartwell-Brat.

Schoeck, H. (1966). *Envy: A theory of social behavior.* New York: Harcourt, Brace & World.

Spicer, B., Emanuel, D., & Powell, M. (1996). *Transforming government enterprises: Managing radical organisational change in deregulated environments* (CIS Policy Monographs 35). St. Leonards, Australia.

Stapley, L. (1996). *The personality of the organisation: A psycho-dynamic explanation of culture and change.* London: Free Association Books.

Weber, M. (1968). *Economy and society: An outline of interpretive sociology.* New York: Bedminster Press.

Willshire, L. (1997). *Psychiatric services: Organising for an impossible task.* Unpublished doctoral dissertation, Monash University, Melbourne.

Winter, R. (1989). *Learning from experience: Principles and practice in action-research.* London: Falmer Press.

# 9

# The Consultant as Container

*Edward B. Klein, Ph.D.*

### BACKGROUND

This chapter describes my experience doing a long-term consultation. The emotional costs of this work suggest that countertransference in long-term consultation needs to be better understood, just as it does in analytically oriented psychotherapy. The chapter presents enough process details for the reader to have a feeling for the type of group process involved in a lengthy organizational consultation, which is becoming less common. Just as managed care pressures have made clinical work shorter, similar economic concerns have led to more time-limited organizational consultations. The following edited chronological account described a 6-year consultation, involving 202 group meetings, to the staff of an inpatient adolescent unit. Note that over time I became a container for staff feelings and thoughts. But, as a container of unit affect and history, I had strong countertransference reactions which affected termination of the consultation.

## THEORETICAL ORIENTATION

My consultation is derived from the infant-mother relational work of Melanie Klein (1959). She noted how early life is marked by two tendencies: splitting one thing into opposites—the good and the bad—which later in life is the basis for stereotypes, and projective identification, which leads to attributing larger-than-life characteristics to leaders. On the other hand, good enough mothering in infancy decreases these tendencies and leads to healthier development in adulthood. This individual theory was extended to groups by Bion (1959). He noted that task teams simultaneously exist at two levels: first, the work group with a formal agenda and time frame; second, the basic assumption mode—including pairing, fight-flight, and dependency toward leadership—which often decreases group productivity.

Rice (1963, 1965), who worked with Bion at the Tavistock Institute, applied individual and group theory to the functioning of complex systems. There are six Tavistock concepts that highlight organizational life. (1) The primary task is the work a unit has to do to survive. (2) A boundary is the region that separates the organization from society. (3) Authority is what managers use to regulate institutional boundaries, or chaos will result. (4) Leaders are "Januslike," looking both inward and outward, allowing them to stand amidst multiple conflicting pressures. (5) Social defenses are unconscious behaviors used to avoid activities that evoke staff anxieties. (6) There are task and social roles; leadership involves integrating these two roles and is a boundary regulatory function (Astrachan, 1970). These concepts are helpful both in organizational consultation and group therapy (E. B. Klein, 1992).

I conceptualize the consultation process as involving five developmental stages. These overlapping phases include (1) external and internal contract negotiation with administration and the group; (2) entering and diagnosing the system; (3) interpreting the overt and covert process; (4) understanding the relationship between the group and the external environment; and (5) spending an appropriate amount of time on the termination process.

The contract is particularly important; it creates a structure for the members regarding attendance and responsibility. It begins the process of building trust, a safe environment, and a division of labor between member and consultant, and helps to establish the external and internal boundaries within which the group is to work. The external boundaries protect and stabilize the work of the group by defining the meeting time, membership, and relationship with the consultant and with others outside of the group. The internal boundaries are the region in which the interactions of the group take place.

## HISTORY

In 1983 and 1984 I supervised Rose*, a former student, in her consultation to the adolescent psychiatry staff of a large general hospital. When I returned from a sabbatical in 1986, Rose had moved to another city. She called to let me know that the service needed a consultant and that I should contact Sally, a Black woman, who was the head nurse. At my request Sally arranged a meeting with the chief psychiatrist, Don, who had started the 16-bed unit. It was a dynamic milieu oriented ward with an average stay of 4-6 months and a waiting list. Don was a bright, charming, liberal man, and a strong advocate of patients' rights. On the other hand, he dominated Sally, and staff seemed very dependent on him. Although he had supported Rose and the process group for years, Don voiced concern about changing from a female to a male consultant, which might be difficult for a primarily female staff.

To provide clarity, the five developmental stages noted earlier will be used to organize the following chronological description of this long-term consultation.

## NEGOTIATING THE CONTRACT

Don and Sally and I negotiated an open-ended contract. I led a weekly group with the understanding that members were to discuss their own dynamics in order to be an effective work team

---
*All names in text are pseudonyms.

and to avoid patients enacting staff conflicts. The group met from 3 to 4 P.M. so that both day and evening shifts could attend. Meetings took place during the academic year, excluding the summer, to accommodate my teaching schedule and Don's budgetary concerns. Attendance was voluntary. Staff was responsible for confidentiality. I felt that discussion of task, boundaries, and roles would lead to a cohesive group environment and that explication of larger systems issues which affected the unit would also be helpful.

## ENTERING AND DIAGNOSING THE SYSTEM

At the first group meeting, on November 4, 1986, the themes were trust and inequality. There were splits by gender, discipline, and race as represented by the seating arrangement—three male nurses on my right, women who were not nurses on my left, and the female nurses and Don opposite me. (This seating arrangement slowly changed over the year as members became more comfortable.) Many women had been on the service since it began. This was a family with a charismatic male head, verbal senior women, and silent young men. I asked people to be on time. Don and Sally were supportive.

Groups during the first half-year centered on Don's authority and anger at his frequent absences. When Don was away members expressed envy of his freedom to travel to distant places. When Don was present other issues, such as tardiness, were central. At the second session members were late. They complained about staff losses, the coming holidays, and lack of support. I suggested that they might use the group for social support. Members constructively addressed this problem by deciding on shorter patient evaluation meetings so that all staff could be on time for the group.

A week later staff anger occurred, aimed at the administration. A Black nurse was disrespectful toward a patient who left against medical advice (AMA) and his mother and was therefore docked 2 days' pay. I interpreted her behavior as an expression of staff desire to behave similarly. Patient-staff relations were openly discussed only after others shared their wishes to behave in the same manner as the docked nurse. In an intensive discussion, including how similar reactions to patients occurred in other hospitals as well, members learned how commonly one person could speak for the group.

In the middle of another session I noted that three women being overweight and one gay man's smoking in the group could be seen as self-destructive behavior. One woman cried, another mentioned her ill child, and the man sadly noted that his lover was leaving town, while Don attacked my moralistic stance. Though there was more sharing in the group, little progress was made on planning for the summer.

During that first year I learned three things. First, negotiating with two senior leaders was not the same as *developing* a contract with the whole staff regarding group goals. Second, external comments about other hospitals are more acceptable than interpretations about staff dynamics. Third, the dependency culture focused on Don was so powerful that my comments on internal group process were not very helpful. Specifically, I had gone beyond the contract focused on organizational dynamics by pointing out how personal life style issues (health and sex) affected the group.

## INTERPRETING THE OVERT AND COVERT PROCESS

In the fall males complained that I should be more myself, rather than play a distant role. Male nurses saw themselves as uncles, while women were viewed as depriving mothers. During that week the hospital opened an adolescent addiction unit; members felt envious and deprived of resources and the recognition that went to the new (and inexperienced) unit. That is, instead of feeling warmth and acknowledgment, staff felt invisible while the new unit bathed in a warm spotlight.

A week later a patient ran away; it was not reported that he never unpacked. Don accused the nursing staff of being passive-aggressive—a very long silence followed. Finally, nurses discussed not being paid time and a half for Christmas as nurses at other hospitals were, thus engaging in a passive-aggressive style of interaction.

In the first group in 1988 Don was on indeterminate leave due to personal problems. Members were depressed during this leadership crisis. Don returned in 2 weeks, and at the next group there was discussion of how to make Don's life easier by senior staff sharing some of his work and having a retreat during the summer. I invited Don to lunch, and he reported that staff pulled him into too many details against his will. He voiced his frustrations with the "cold" hospital administration and "dependent" unit staff. I sided with him against the administration. My moving closer to Don made it difficult to maintain distance in my consultant role to the whole unit.

In the following month staff were fighting with patients who were staying for shorter periods, since this entails less treatment and fewer staff rewards in terms of patient improvement. With shorter stays there were fewer work leaders among the patients, which also contributed to staff frustration. Members wanted Don to find a place for the retreat; he resisted, suggesting that they be more proactive themselves and more understanding of patients. Nevertheless, he used his faculty status to obtain a place for the retreat.

The all-day summer retreat finally happened, and 24 out of 26 staff attended. The design was morning diagnosis and afternoon planning sessions. Members formed groups on Active Treatment versus Custodial Care, Staff Morale, Interdisciplinary Communication Problems, and Transference/Countertransference Issues. At the next session staff excitedly reviewed the retreat and planned small groups every 2 weeks and community meetings monthly during the 3-month summer break when I was away. Don voiced doubts about the new plans; as members became more independent, he became more controlling.

In the fall of 1988 there were angry reports of runaways, frustration with the administration, and an acknowledgment of the mutual dependency between patients and staff. There was an awareness that, with shorter patient stays, staff needed to obtain more support from each other. As an illustration, when a women reported that her brother had died, I talked about my mother's recent death, which brought us closer as a group.

The following week a dynamic nurse administrator joined the group, and we heard Sally's belated report from the retreat group on Active Treatment versus Custodial Care. At the next session there was a report from the Interdisciplinary Communication Problems group. Staff worked to integrate retreat reports into current unit dynamics. This was followed by a fight between Don and Nursing; nurses feared losing control of their work schedules to outside administrators, and Don called them "cry babies."

At the first 1989 session, with Sally's leadership, Black and White staff collaboratively discussed the importance of not meeting on Martin Luther King's birthday, in order to honor his life and unfinished work. The next three groups were marked by complaints about nonrewarding patients, while nurses voiced fears of a slow summer with possible layoffs. After discussion they acknowledged that these concerns were linked, and members stopped displacing negative feelings toward the administration onto patients.

In the fall the issues were a strike, more patients on the unit for less time, and no waiting list. After a number of these meetings, members adaptively decided that they needed to speed up diagnosis and care in order to keep the unit viable and protect their jobs. That is, staff realized that they had some control over their clinical and administrative tasks that positively affected the work environment.

In 1990 there were sicker patients, fewer staff, and many new admissions, leading to more stress. One session focused on staff secrets (dating among staff members and relations with former patients), stimulated by attendance at a women's workshop. Don and a woman battled over gender issues, while others tried to avoid acknowledging both senior staff members' fighting in public as well as serious systems issues.

In the fall there was a cutback of middle-level managers, including the dynamic nurse administrator who had been in the group for a year. Nurses were in mourning. The assistant head nurse left to work in a well-baby clinic. The following month a new administrative nurse joined the group to learn about the unit. Her son was in the Army, stationed in the Persian Gulf. Even with loss and threats of violence against the nurse's son in the Gulf, there was a Christmas party that I attended, and I felt more part of the "family."

In terms of learning, it was clear that I contained more of the group affect and identified strongly with staff. Also, even with a contract there were control issues with Don, particularly when staff acted independently. Although I was more direct than Rose, Don's competitiveness with me did decrease over time. Most importantly, even with system distractions member learning occurred, as evidenced by greater awareness of group dynamics and more cooperation (e.g., earlier diagnosis and treatment planning) on the service.

## UNDERSTANDING THE RELATIONSHIP BETWEEN THE GROUP AND THE ENVIRONMENT

In 1991, in response to changing economic circumstances, the hospital declared the ward a short-stay unit. The Gulf War started, and the administrative nurse's focus on the war (her son was in the Army) made it difficult to explore unit dynamics. Patients engaged in violent acts, reflecting the Gulf War. Staff, encumbered with sicker patients, were challenged to support each other. Group attendance dropped; with fewer patients, a general mood of depression prevailed. For the first time the census dropped below five; there were "voluntary" vacations and anger at the administration for failing to obtain new patients. Staff were into a survival stance, with war as the metaphor for unit dynamics. I held three extra sessions, which Don paid for, to meet pressing group needs.

In the first fall group session we learned that Don and Sally were leaving, he to head a community mental health center in California and she to start a partial hospitalization service. The next seven sessions were rich with history and feelings focused on Don's leaving. A White male nurse was appointed to replace Sally. A woman psychiatrist, who had been with the unit for years, was the staff choice to replace Don. She was offered an *acting* position, was insulted by the offer, and therefore refused it. A male referring psychiatrist, Lou, accepted the position as acting head with the understanding that there would be a national search for a permanent head. Don left; although the situation was still unsettled, Lou started building good relations with staff. There was a sense of hope, since Lou was responsive to both the administration and the group.

In 1992, after he left, attitudes toward Don became openly negative. Staff had both transferential feelings toward and dependence on Don. They wanted a good father but sometimes got a rageful one. Each fall he discussed leaving to work elsewhere. Don's threats of leaving made it difficult to directly address him. At the next session there were group feelings of loss and belated mourning for Sally. The senior educator left for a job in Chicago. Three of the original five senior staff had left within 2 months. Even with these losses staff sustained a sense of hope in Lou's emerging leadership style. For the first time in 5 years I missed a meeting, possibly an unconscious reaction to the changing situation.

During the next session it was announced that there were three candidates for the chief position and that staff would have input in the selection process, which pleased members since they wanted Lou. At the next meeting it was reported that there was only one candidate for the chief position. Lou said that he was not the candidate. The group despaired that staff wishes had again been ignored. The next week reports circulated that a nurse had money stolen from her purse and that a male patient had threatened a female staffmember. It seemed that the uncertain leadership situation produced more violence (stealing and physical threats) toward women staff.

Two weeks later an outside psychiatrist, Henry, was appointed as the new director. The staff was depressed and withdrawn, as they had not been consulted about this appointment. Over the following weeks patients acted out, videos were stolen, layoffs occurred, the woman psychiatrist left, and meetings were marked by staff's crying. Meanwhile, Henry articulately presented managed care as the future direction for the unit and demonstrated his knowledge of external boundary issues in the larger health care environment.

In the fall of 1992 meetings started late; staff racial issues resurfaced; patients left AMA, engaged in violent acts, and set fires; and female staff felt unsafe. A male and female nurse fought and then realized that they needed each other. Staff members felt that the administration was distant, uninterested in patient or staff needs. There were problems with both referring and unit doctors; the system seemed to be in flight and falling apart. On the anniversary of Don's departure a staff member remarked, "We have gone from being the Maisonette (Cincinnati's five-star restaurant) to the McDonalds of mental health!"

## SPENDING TIME ON THE TERMINATION PROCESS

Before the next session Henry said that due to economic difficulties the consultation had to stop. I was hurt. The next two meetings were sad, involving some discussion of my role in the group. In the last session I had more feelings than anyone else acknowledged. There was a possibility that one of the nurses, who was on another service, would run the group in the future.

After I left, the off-unit nurse ran the group. Only nursing staff attended, reflecting a year-long trend, identification by discipline rather than service. This was one way of coping with an administration whose primary task seemed to be the bottom line, not quality patient care. Not surprisingly, the group ended after 2 months. Because of a lack of patients, the service was forced to close a number of times in 1993. In 1994 RNs on the addiction unit were laid off and nurses from the adolescent service had to cover on that unit. The addiction, partial hospitalization, and adolescent units were integrated in order to produce a more effective system. Many staff felt deceived and angry and wished that the group would be restored as a public vehicle for discussion of common concerns.

In sum, over the course of 6 years the major issues were trust, dependency on Don, gender and racial relations, the loss of senior leaders and middle managers, insurance policy changes resulting in briefer treatment with more violent and resistant patients, and greater use of medication. These systematic changes led to a sequence of low morale, Don's departure, an acting director who inspired hope, and an externally oriented director who ended the consultation.

## DISCUSSION

The external and internal aspects of the contract were effectively negotiated; entry and diagnosis were well done; overt and covert group processes were competently interpreted and systems issues were analyzed so that staff felt knowledgeable. For example, I encouraged Sally and Don to set good examples by being on time and speaking up. Appropriate boundaries were drawn around the group, creating a safe environment in which to examine difficult personal and ward issues. We explored other clinical settings, which normalized many internal staff concerns. The cooperative atmosphere allowed for a more realistic appraisal and credit giving as well as criticism.

As long as Don was the director, the group worked well as a team, since he supported the external boundary to the hospital administration. With a strong group boundary, open discussion was possible of splits by gender, race and discipline, projections onto individuals and subgroups, and hospital and unit leadership; thus, members felt like part of a functioning team. That was true for 4 years with Rose as well as for the first 5 years that I consulted to the group. For 4 months with an acting director who accepted the importance of staff processing, it was a hopeful, workable group. When the new director with his managed care perspective took over it was just a matter of time before the group ended. The stated reason was finances, but of equal concern was the fact that the group represented the past, a reflective space to think together. I may have symbolized the warm and more fulfilling past versus the economically driven present. Most importantly, the management philosophy in health delivery systems was no longer in harmony with a staff process group that could question the direction of the administration and unit leadership.

With three leadership shifts in 1 year and little staff input into decision making there was uncontained change marked first by hope and then by flight, lack of trust, alienation, and cynicism. I did not renegotiate the contact with Henry to reflect the new direction the unit was taking as a brief treatment unit. Possibly this was because I did not wish to pay deference to Henry, nor did I expect him to be supportive of my work. He had never bought into the contract and saw little need for a staff process group. Therefore unresolved feelings probably interfered with my role performance throughout a period of significant change. For example, I could have fought harder for a few more termination sessions, which might have been useful in terms of understanding the process and thus providing some form of closure.

In retrospect, a shadow consultant, who could have helped me to see more clearly what was occurring, would have been very useful—particularly in the year after Don left. This would have aided me in not internalizing the process and in analyzing my only missed session as an unconscious awareness of the changed situation. Nevertheless, the larger systems issues realistically ended the group, regardless of my own feelings. Recent discussions with the male head nurse and Sally reinforced the notion of the power of the changed management philosophy as being the overwhelming factor in the termination of the group.

Yalom (1985) noted that patients join a therapy group with existential anxieties about loss and death. From an object relations perspective, terminations provoke a grief reaction. Terminations in larger work settings lead to the group's avoiding the depressive awareness of loss. Indeed, an ending can be experienced as a boundary violation. Terminations disrupt the illusion of the all-giving, supportive, maternal leader and the permanence of the group. The unit was, in fact, disorganized at times and, consequently, group members felt abandoned. Because patients' treatment was significantly shortened or terminated by new insurance policies, staff left, threatening the unit's very existence. In response, adolescents frequently acted out in violent fashion. Organizations often respond to loss by using social defenses such as avoidance and denial. The staff group's task was to understand these forces and try to respond in an adaptive fashion; at times, over the first 5 years, they did so with great creativity.

The central role of projective identification in treatment groups has been noted by Ashbach and Schermer (1987). At termination a single person can become a container for the projections of group members. As a container filled with others' affect, I missed using termination as an opportunity to help the group integrate these projections through proper collective containment. That is, I should have helped each member to own his or her own thoughts and feelings so the process would become a group rather than individual issue. This, I might add, was a difficult task with Henry in the room and members worried about their jobs.

In the development of the object relations approach, time boundaries have influenced theory and practice (Miller, 1959). All relations and groups are time-limited. In this long-term group the time element was often denied. In our usual time-limited organizational work the reality of time boundaries speeds up the consultation. An extended intervention provides depth of understanding and opportunity to work on long-term issues but may lead to a focus on internal issues rather than on the greater impact of external economic and social forces influencing the institution in which the group is located. Objectively, it was the power of changing external mental health policies that terminated the group more than any individual or group dynamic.

Two central object relations concepts are basic assumptions and leadership. Although Don's leadership fostered dependency, members worked on their group dynamics, which facilitated humane patient care and staff-patient interactions. Lou's collaborative leadership style was marked by the hope that staff would be listened to and have a voice in the unit decision-making process. Henry's leadership led to members' flight into their discipline, lack of identification with the service, and departures from the unit. At times members engaged in sophisticated pairing with me on the primary task, which demonstrates how complex group dynamics powerfully affected patient care and staff-patient relations. Over the years I became the voice of the good, dependable past, rather than the changing, unpredictable present. Thus, as the container of the past, I had to be terminated so that new leadership could build a cost-effective, future-oriented unit.

Consultants need organizational members just as teachers need students, to do the work. There is mutual dependency and vulnerability. For example, when a member lost a relative I shared my experience of the loss of my mother. Termination heightens these feelings of grief. For instance, if I had had other consultations the termination of this long-term hospital work would have been less personally disruptive. In some ways this chapter is part of my grieving the loss of this compelling 6-year consultation and the quality care the unit used to symbolize.

Termination was inadequately processed because of my overidentification with the unit. I had become a container for the group's affect, and consequently lost my systems perspective. I was unable to perceive that changes in the mental health field rather than personal or professional shortcomings forced me out. I did not manage myself in role (Lawrence, 1979); that is, I understood intellectually but was caught up in feelings of abandonment which, of course, paralleled what the staff was going through at that very moment.

In addition, there is an interplay between adult development and gender involved in this consultation. Just as women are carriers of the story (Gabelnick, 1998; Gold, 1998) as the container of unexpressed feelings and history, I represented the feminine aspect of the unit. Terminating the process consultant meant focusing exclusively on a male, hard-work management philosophy. Also, my own post-midlife stage of adult development stresses more of a balance between the masculine and feminine, which does not fit into a male, bottom-line, market-driven organization.

Consultants have strong power, as well as financial and self-esteem, motives for maintaining the group. Consultants also have a need to cure the "sick" organization. Unconsciously the consultant may attribute an earlier-than-planned termination to a personal failure, professional incompetency, or unresolved grief. The reality of the loss adds to the countertransference so that the consultant cannot see his or her distortions, as is possible when the group is in a more stable stage.

On a societal level more anger emerges as formerly good enough work settings no longer are safe containers for people's career aspirations and hopes. With the breakdown of the psychological contract between organization and individual, there has been a marked shift into professional rather than institutional identification (Astrachan & Astrachan, Chapter 2, this book). This anger is expressed in the workplace and in the larger community through violence (destruction of a federal office building and an Amtrak train) and social-political disaffiliation (record low voter turnout in 1994). It may be no accident that in a 6-year consultancy, I, as the outsider, became the container of group history and feelings in a unit that no longer provided a safe container for personal expression. I became the insider who honored the need to process what happened in the organization. This may be one danger inherent in doing long-term consultation alone in an unpredictable environment with ever more permeable boundaries. As the health care system and the unit moved from security and fulfillment to unpredictability, members got "rightsized" (i.e., downsized). Today, consultants may also have to be seen as interchangeable, identified with one relatively quick approach (i.e., reengineering, continuous improvement, matrix management), in order to be like other group members who also have to be efficient in an ever-changing world.

# REFERENCES

Ashbach, C., & Schermer, V. L. (1987). *Objective relations, the self, and the group: A conceptual paradigm.* London: Routledge & Kegan Paul.

Astrachan, B. M. (1970). Towards a social system model of therapeutic groups. *Social Psychiatry, 5,* 110–119.

Bion, W. (1959). *Experiences in groups.* London: Tavistock.

Gabelnick, F. (1998). The myths we lead by. In E. B. Klein, F. Gabelnick, & D. Herr (Eds.), *The psychodynamics of leadership* (pp. 323–345). Madison, CT: Psychosocial Press.

Gold, U. (1998). Feeling the work in one's bones. In E. B. Klein, F. Gabelnick, & D. Herr (Eds.), *The psychodynamics of leadership* (pp. 297–321). Madison, CT: Psychosocial Press.

Klein, E. B. (1992). Contributions from social systems theory. In R. H. Klein, H. S. Bernard, & D. L. Singer (Eds.), *Handbook of contemporary group psychotherapy: Contributions from object relations, self psychology and social systems theories* (pp. 87–123). Madison, CT: International Universities Press.

Klein, M. (1959). Our adult world and its roots in infancy. *Human Relations, 12,* 291–303.

Lawrence, W. G. (1979). A concept for today: The management of oneself in role. In W. G. Lawrence (Ed.), *Exploring individual and organizational boundaries: A Tavistock open systems approach* (pp. 235–249). London: Wiley.

Miller, E. J. (1959). Technology, territory, and time: The internal differentiation of complex production systems. *Human Relations, 12,* 242–272.

Rice, A. K. (1963). *The enterprise and its environment.* London: Tavistock.

Rice, A. K. (1965). *Learning for leadership.* London: Tavistock.

Yalom, I. D. (1985). *The theory and practice of group psychotherapy* (3rd ed.). New York: Basic Books.

# Part II

## Consultation in Health Settings

# 10

# Downsizing and the Accidental Consultant

*Peter Herr, M.A.*

Perhaps one of the most ubiquitous business decisions faced by corporations is whether to downsize, or, euphemistically, rightsize, in order to remain competitive. While there is no question that corporations must remain fiscally sound to remain competitive, downsizing is not without its own costs, many of them unanticipated. One field in particular, health care, has experienced major changes and contractions in its work force to accommodate the changes required by health maintenance organizations (HMOs), preferred provider organizations (PPOs), and reduced fees for services. These recent drastic changes are due, in part, to health care's previous freedom from normal business pressures and competitiveness. Increased costs were passed on to insurance carriers and consumers with little regard to cutting costs or finding more efficient means of operating. However, the effects of downsizing in health care are instructive to other businesses because issues of leadership, communication, and employee satisfaction play important roles in any business operation. Unfortunately, these issues often go unaddressed or are ignored during a downsizing until such time as they become a crisis, yet the impact of these factors will often determine the organization's success or failure.

## LEADERSHIP AND THE ORGANIZATION

Krantz (1997), who bases his contention on Bion's work (1961), states that the emotional health of an organization can give it a competitive advantage. We are currently living in a period of unprecedented rapid change and vulnerability. This leads to greater insecurity and fear of isolation and annihilation among the corporation's employees, and so greatly endangers the organization's emotional health.

Traditionally, organizations met individuals' dependency needs through their structure, processes, and containment function. A well-functioning organization might operate at a depressive level, which Krantz conceptualizes as being grounded in reality and having a greater likelihood of success. However, bureaucracies are shrinking, processes are constantly being transformed, and the organization is no longer seen as a safe container. Fear of takeovers, layoffs, or job shifting has resulted in many corporations functioning at the paranoid-schizoid level, in which employees, including management, are emotionally and intellectually disabled; there is little collaboration between groups and a major organizational focus is scapegoating. Employees cannot link their work to the wider organization's purpose and have a difficult, conflict-filled relationship with the company hierarchy. Anxiety has made them more defensive and less likely to work together. When a downsizing does occur, decreased morale and employee commitment are the result. Often the "survivors" are left in a regressed and immobilized state.

The irony, of course, is that while there are increasing external pressures to compete (which multiplies the probability of a paranoid-schizoid stance), there has never been a greater need for confident, capable, and invested employees able to work together productively. Fortunately, we now have a well-educated work force; unfortunately, it is still rarely integrated into decision making. In the past labor and thought were split. Workers were not expected to contribute anything besides their labor. For many managers, this dichotomy remains an ingrained way of doing business, and is an incredible waste of talent.

While it would be easy to lay the responsibility for increased worker anxiety and dissatisfaction mainly on poor leadership and management, there are other significant systems issues. Miller (1998) contends that management is increasingly segregated and emotionally disconnected from the workers they manage. As vast numbers of middle managers have been "delayered," the corporate pyramid has changed shape. It is a much steeper pitch from line worker to CEO. The middle manager who used to serve as a bridge and a buffer has assumed the responsibility of several peers and is overwhelmed by increased role requirements and anxiety over anticipated cuts and external competition.

Upper management is often seen as a hired gun; they are usually brought in from outside, as opposed to promoted from within, to perform some short-term changes. They rarely make a commitment to stay beyond 3-5 years. These managers come in to clean up a mess through "reorganization," which brings about some short-term gains. However, when the long-term consequences of their short-term strategies appear, they already have prepared their résumé, or are long gone. Those long-term employees who carry a sense of the organization's history are well aware of the pattern. While the stockholders may be happy for the moment, the employees become cynical and less loyal. Miller suggests that in order to lead the troops and deal with their own needs for security and certainty, management requires a common enemy. As customers are increasingly the focus of service, and today's competitor may be tomorrow's employer, management picks its own employees as the surrogate enemy. After all, they consume an inordinate amount of resources in wages and benefits. If they can be adequately cut the enterprise will survive and stockholders will be satisfied. The methods are outsourcing, use of temporary employees, and transfer from full-time to permanent part-time employment.

The workers are not blind to this scapegoating. When employees at all levels perceive that they are disposable, they may chose to be compliant, but they have also become more calculating. The job is a paycheck, another means to an end. Belonging to an organization may have become a quaint notion, loyalty a thing of the past, and overt cynicism the tune on employees' lips. Leaders and top management become the objects of rage or contempt, or are disregarded and marginalized.

Miller (1998) asserts that organizations are important to people because they confer identity and give meaning: a surrogate family. Institutions allow the expression of primitive emotion and provide defenses against anxiety. When the organization no longer provides these defenses the employees psychologically withdraw. If workers feel attacked by the very institution that is supposed to protect them they are left in a fight-or-flight stance. Given the unequal power distribution, fight is rarely a viable option. Flight, particularly in an economy with very low unemployment, becomes an opportunity to advance oneself and to punish the employer.

Management has two reactions to this psychological withdrawal: deny its existence or try to increase commitment through reindoctrination. Many leaders either believe that downsizing has relatively little effect on motivation, or that they can increase commitment through incentive programs, company videos, mission statements, and total quality management. These efforts often create the opposite of the intended effect: The employees may find the efforts laughable, and disenchantment, resentment, and distrust increase.

Miller suggests that managers see the staff either as programmed robots without feelings or as infants who will believe whatever their "parents" tell them. This represents two styles: those of a noncaring leadership, and those who believe that they can be the employees' ego-ideal.

## "THE CONSULTANT"

I had the opportunity to observe the dynamics described herein at a medium-size hospital in Indiana where I was employed in a nonconsulting capacity. I had worked for the hospital as a part-time therapist for over a year when I was asked to apply for the department's Unit Manager position, a role I held for 18 months, until the department was permanently closed. The Vice President (VP) of Patient Services then asked me to assume the Patient Advocate's position until I began my clinical internship 18 months later. Both the VP and my direct supervisor, the Patient Services Manager (PSM), were agreeable to this arrangement and believed that my range of experiences would be a valuable asset to this relatively new position. My duties were to include counseling in various departments (Emergency Room, Oncology, Women's Center, etc.) along with handling patient concerns and complaints throughout the hospital, although primarily on the three main medical floors. I would also handle various other tasks as directed by the VP and my supervisor.

This position allowed me to speak to, and often to get to know rather well, many of the personnel from housekeeping up to the VP. In my previous position I hardly knew any of the employees because our department was off site, with only minimal contact with the main hospital.

My past interest in organization and social system dynamics was stimulated by the transitions that were occurring at work. Prior to taking the Patient Advocate position I had helped to edit this volume and also its companion on leadership (Klein, Gabelnick, & Herr, 1998), along with taking courses in Organizational Development. However, other than my previous supervisor, no one at the hospital was aware of this background. While my knowledge of organizational dynamics was helpful in understanding the process when my department was closed, these skills now seemed largely irrelevant.

This remained true for the first 6 months. At that time the VP of Patient Services resigned after being in this role for approximately 5 years. Previously she had worked as a hospital consultant and was resuming this career path. Her resignation gave me the unique opportunity to see the effects of a leadership transition and downsizing through the eyes of an employee with knowledge of organizational dynamics who had developed an open and supportive relationship with a large number of staff members. I was privy to the staff's thoughts and feelings in a way an outside consultant, with limited time and resources, could never hope to be. On the other hand, these relationships obviously compromised any claim for impartiality and objectivity. While my stance might not be completely neutral, I was still an outsider in that I was neither a line employee (nurse, aide, etc.) nor a permanent professional. The majority of staff knew my relationship with the hospital would be relatively short. Further, they welcomed my empathetic ear and trusted that their conversations would remain confidential, because of my training as a therapist and because I was seen as outside the chain of command and not as a part of management.

## "THE CONSULTATION"

### The Transition

The new VP was always pleasant, cordial, and respectful to me and my clinical work. She had lived in my hometown, and we found other matters of mutual interest. She has a good sense of humor, and our biweekly meetings were productive and enjoyable. I had mixed feelings about her predecessor; she had treated me well, but had closed a well-respected department, causing considerable stress to me and my coworkers.

In getting to know the hospital staff it was clear that the former VP was well liked and respected even though she had made numerous mistakes and caused considerable stress when she first arrived. Employees felt she was warm and open and willing to listen to their concerns. Shortly after her arrival she made numerous changes, which included downsizing on several floors. This resulted in a large turnover, many problems, and the eventual return to previous staffing levels. In addition, nurses began working in teams, and on two of the floors a system of charge nurses, who shared management responsibility, was instituted.

The hiring of a new VP roused considerable staff anxiety as they clearly anticipated a replay of previous mistakes. Several employees told me how this was their third, fourth or fifth VP transition and how they anticipated a minimum of 1 to 2 years of severe problems brought on by the "new broom sweeping clean." If the VP did an adequate job of adjusting to the circumstances, there would be a short period of stability followed by another transition and a round of chaos as this VP left for greener pastures. At worst, the new VP would do a poor job, and there would be ongoing instability until an adequate replacement was found.

The arrival of the new VP did little to lower anxiety. On her initial tour of the hospital she appeared stiff and somewhat authoritarian. Most employees continued to hope for the best.

Immediately the staff began to compare the two VPs. The former VP had her office on one of the medical floors and her door to the main corridor was almost always open, even though there was considerable traffic. Staff felt that they could approach her at any time and discuss their concerns. The new VP kept this door closed, and any visitors were required to go into the main office and check with her secretary. With time, the new VP had her office moved off the medical floor and onto the administrative floor near the chief operating officer (COO) and CEO. While the reason was ostensibly that it was more convenient for her to attend meetings, employees saw it as an attempt to distance herself from them. This was consistent with staff's belief that she did not want to get close to them, either because she did not care to know what they thought, or did not want to get to know those she was about to terminate.

Staff anxiety was escalated further by the conflict between the new VP and the PSM. The PSM had been closely allied with the previous VP and the latter's resignation was a significant loss, as the VP was both confidant and mentor to the PSM. The new VP and PSM seemed to be in conflict from the start, and this was acted out symbolically. The PSM had only been four doors away from the old VP. The new VP implemented plans to move the

PSM "closer to the charge nurses." In doing so, the VP and PSM would be separated by the height and width of the building. The PSM resigned for a position at another hospital before this change was implemented.

## The Downsizing

The conflict between these two was in part due to the VP's belief that the hospital was overstaffed and needed to be downsized. Attempts by the PSM to counter this perception were met with stiff resistance. The VP would hear none of it, and further attempts by the PSM resulted in her feeling frustrated, powerless, and isolated. She feared further attempts to persuade the VP would only result in greater animosity and eventual termination.

When the downsizing occurred it was far worse than the staff had expected. On two of the hospital's three main medical/surgical floors, staffing was to be reduced by one third. The hospital has a policy of never laying anyone off. Employees were simply asked not to report as previously scheduled when the patient-staff ratio indicated they were not needed. Staff quickly ran through banked vacation time accumulated over years to supplement their reduced pay. Rumors ran rampant: For example, the hospital was being downsized so it would look more attractive to a potential buyer; it was losing money and all nonessential personnel were going to be eliminated. While these rumors were repeatedly denied, their persistence speaks to the staff's anxiety and vulnerability.

The downsizing was instituted because the VP wanted to reduce expenses to a level closer to a national average, even though the hospital was somewhat unique in that its clients were primarily nursing home geriatric patients, many of them requiring total care. She also wanted to eliminate most of the LPNs. Many of these experienced nurses had been employed by the hospital for 10 to 25 years. Her plan was to replace them with RNs, as has occurred in most of the hospital industry.

This is not what happened. The LPNs went through their paid time-off benefits, but stayed with the hospital. The RNs and the charge nurses (the hospital's first line of management) left in droves. On one floor, all three of the charge nurses quit or transferred to other jobs outside the hospital, but within the corporation. On the other floor, one charge nurse quit, and the other two stated they would rather return to direct patient care than take on full management responsibility as the VP suggested. (One eventually returned to direct patient care while the other remarked that this was her third transition and she would survive this one as well.) Of the regular RNs, a large majority either transferred or quit. Before the new patient-staff ratio was announced there had been an excess of both RNs and LPNs because efforts had been made to ensure adequate staffing through the high census winter and spring months. Five months after the downsizing there were insufficient nurses of either kind to staff at the new lower level. It was at this point the VP addressed the LPNs about their importance and how they had carried the hospital through past nursing shortages. They were unimpressed by her sudden change of heart.

Conditions continued to deteriorate. Nursing staff was asked to work overtime at time-and-a-half pay. They refused, stating they were too exhausted at the end of their shift to work any overtime. Appropriate staffing levels were maintained for a period by offering double-time for those nurses willing to work. Ultimately, even this was an inadequate incentive and outside agency help was required, at considerable cost to the hospital. The VP attempted to reassure staff at this point that help was on the way in the form of new hires. However, the employees were well aware that bringing new hires on board could take 2-3 months and another 4-6 months until they were adequately oriented and trained. This bleak scenario assumed new hires were retained in the face of the demanding workload. Human Resources resorted to offering a recruitment bonus to any employee who was able to entice an RN to join the hospital.

Consequences of the downsizing were not limited to these two floors. Nurses from other units were pulled to these floors when there were shortages. While a certain amount of grousing might be expected when people are taken out of their routine, these nurses became resistant and angry. Nurses referred to it as being shipped to "Vietnam." In addition, a new unit authorized to have a much higher patient-staff ratio opened shortly before the downsizing. Because the workload of staff on that unit was generally perceived as much lighter, and they were largely exempt from being pulled to the other floors, the staff elsewhere were resentful and envious.

## Scapegoating

Before the downsizing the nursing staff worked hard and did an excellent job. The staff supported each other, had a positive attitude, and took pride in their work. With the downsizing employees regressed and become immobilized. Part of this was due to role overload. Even before the downsizing, departing staff were not replaced and duties were distributed or dropped entirely. The staff assumed a survivor mentality, hoping to make it through the week. I heard repeated employee complaints about being exhausted and unable to work adequately, as well as worries that sooner or later something would be missed and they would be blamed or lose their license. The stressed staff would joke that instead of listening to patient complaints, I should be assigned to hear theirs. Fortunately, employees did not act out their frustration by aggressing against the patients. A significant rise in complaints, starting with the first month of the downsizing, did not include complaints about rude or abusive nurses, but did include complaints about poor response time. Previously, this had not been a significant problem.

The VP was expecting some turnover because of the downsizing and had prepared her arguments. The fault was clearly laid at the feet of the nurses or those who hired them. Because the hospital was relatively small and had limited opportunities either for advancement or training in nongeriatric care, it had a pattern of hiring recently graduated nurses, training them, and then seeing them leave after 12-18 months. The VP assured me a month after the downsizing that any turnover in the near future would be due to the large number of nurses who were approaching 18 months with the hospital. Conversely, if nurses did not leave, things really were not that bad to begin with. When I suggested that the nurses felt unable to adequately perform their duties, she indicated that they never feel they are doing enough for patients unless they are on a one-to-one ratio. Further lack of concern about staff turnover was evidenced by her elimination of the nurse retention committee.

The VP was correct in that quite a few nurses with 12-18 months' experience did leave. But so did nurses with 3, 5, and more years' experience. These were often young, energetic, hardworking, motivated RNs who had no wish to leave. They were quickly snapped up by other departments in the corporation or by other area hospitals. None reported having any difficulty finding work; most were actively pursued.

Once the exodus began in earnest, the VP argued that these employees really were not committed to this type of work and that they had hired on "just to get the experience." They did not know how to work hard, and did not want to. They had it too easy here "compared to other hospitals," and it was fine to let them go; they would see that it is not better elsewhere.

The VP blamed the PSM and Human Resources for having hired this type of RN and then structuring their work in unrealistic ways. The hospital should not have employed these new graduates; instead they should have hired experienced people who have worked with this population. The PSM and Human Resources should have foreseen this because it had happened repeatedly in the past. However, what the VP ignored was that the hospital, being smaller and independent, could not offer the salary, benefits, and types of clinical experience that would attract nurses in the same way as its larger competitors.

## Conflict

As the hemorrhaging continued and serious staff shortages began to occur, the VP belatedly did become concerned. She was well aware of being savaged in the exit interviews conducted by Human Resources. With a good deal of rationalization she proclaimed that she looked forward to a planned employee attitude survey so a satisfaction baseline could be established and she could discover the problems, other than staffing, that concerned the nurses.

Perhaps one of the most telling moments occurred when she questioned me about whether the nurses were considering unionizing. Her concern was understandable as two other area hospitals were well along the way to being unionized and a third hospital, one of the largest, had just received word that the nurses wanted to organize. It also showed her limited understanding of organizational dynamics and institutional history. The hospital was really too small to be of much interest to the union, the staff was too conservative, and there had been no previous call for unionization during other difficult periods.

While this situation was enough to give most managers headaches, it was really much more serious. The hospital was scheduled to be surveyed by the Joint Commission on Accreditation of Health Care Organizations (JCAHO). Announcement of a survey is usually guaranteed to send most hospital administrations into crisis and/or panic mode. Loss of accreditation ultimately causes the hospital to close its doors. While this is rarely done, failure to meet standards results in more frequent visits by JCAHO to ensure its recommendations are being implemented. At the hospital, a mock survey was held by hired consultants and numerous hours spent on updating policies and procedures. As the clock ticked down to the survey, the VP watched as her employees continued to leave and morale plummeted. A poor review would be cause for termination, as the majority of services to be surveyed were under her direction.

Staff attitudes were unlikely to be helpful during JCAHO review. The relationship between the VP and the staff on the two floors had become so antagonistic that there was little room for compromise or cooperation. Perceptions were clouded with distrust. Staff started to abuse the system by working overtime for double-time and then calling off the next day when they would have received just their regular pay. On a particularly busy day with inadequate staffing, the VP provided sandwiches for the staff. One of the employees pointed out that the number of sandwiches exactly matched the number of staff working that shift. What had been a supportive gesture was seen as another example of tight-fistedness. Nurses suggested the VP come up, put on some scrubs, and clean bedpans. They were resentful of her making rounds and asking the charge nurses how things were going, but not taking their concerns seriously. Some took to working while the VP talked to them. When staff heard there might be an attitude survey, they discounted it; past surveys resulted in no visible changes. It was interpreted as another attempt by administration to show they really cared when they actually did not.

As time progressed, the employees were not the only ones angry and petulant. The VP called a meeting so employees could air their grievances and hear her plans for dealing with the staffing shortage. No one came. When she asked why, she was told a mandatory in-service had been scheduled for the same time and the floors could not spare anyone else. Her reaction: "I guess no one thought my meeting was very important." This example is instructive of how little communication existed within the entire system. She did not check with the charge nurses, and none of them bothered to tell her about the conflict.

One of the most vehement staff reactions stemmed from an E-mail sent to the nurses by the VP instituting a policy where the day and afternoon shifts would help cover the night shift; once a month each nurse would have to work one night shift. On the face of it, it seemed fairly reasonable; other hospitals require their staff to work shifting schedules. It created a firestorm. Several nurses approached me the next morning to ask if I knew about the E-mail. They were furious and were absolutely opposed to working even one night a month. Some had legitimate reasons, including child care concerns and other responsibilities. Several said that the day they were asked to do this they would walk out whether they had another job or not. Others said they would call in sick and they did not care if they got fired. The VP response to these reactions, which I passed on to her, was to say that none of the staff had been hired solely for one shift and this was common practice at other hospitals; it was not asking too much of employees. As staff continued to leave, this policy was never instituted, although new hires were required to comply.

## The Nonconsultant's Consultation

I began meeting with the VP a month postdownsizing after I had sent her several memos outlining the sharp increase in patient complaints. If she had a motive outside of her concern for the quality of patient care, I am unaware of it. Initially, discussions were very difficult. A return to the previous staffing level was not an option—end of discussion. As her concerns about staffing and JCAHO increased, they became more evident in our discussions. With time she saw me as a way to tap into employee attitudes. I never sugarcoated my remarks, as I was distressed by what I perceived as a decrease in the quality of care and increase in staff burnout. I was able to tell her that the employees resented her "walk throughs" and that her questions were not seen as part of a genuine concern for them. She was perceived as unfair, even when she asked the staff to make minor sacrifices.

It was about 4 months after the cuts and the serious problems had started that I informed the VP that I had organizational development background and was familiar with leadership issues. She had become more willing to listen, although she only made some half-hearted attempts to work with the staff. She did meet with the employees shortly after her unsuccessful first attempt. Their perception of the meeting was that new employees were on the way and, according to the VP, that would take care of the staffing shortage. There was no consideration of returning to a more favorable patient-staff ratio. Her attempts to lighten their workload through reducing some of their assessment responsibilities was seen as only having minimal impact. Six months after the cuts, Human Resources found it necessary to again hire newly graduated nurses, while several of the new hires found the work too stressful and quit.

The VP began meeting with several nurses individually, some of whom were able to directly express concerns about morale and how the VP could improve it. The VP even acknowledged to me the need to get staff involved in decisions, "so they would buy into it." In the same conversation, she proceeded to talk about how certain employees with bad attitudes were infecting the new hires; this behavior was not going to be tolerated, but would be dealt with firmly.

My efforts came to a conclusion with the start of my clinical internship. Unfortunately, my major achievements appear to have been raising the VP's awareness of employee dissatisfaction and providing a container for the staff's affect as they struggled with their own unfulfilled needs and anxiety. The VP was still unwilling to genuinely open communications with her staff, seek input, and implement suggestions. The JCAHO was expected the month following my resignation.

## ANALYSIS

The need to reduce costs is understandable, particularly in light of the squeeze applied by managed care and reduced reimbursement. However, the VP may have saved money by reducing her patient-staff ratio only to see her costs for overtime, agency help, and attracting, hiring, and training new employees go through the ceiling. And how do you calculate the cost of the loss of frontline managers, lack of effective communication throughout the system, and greater staff and patient dissatisfaction?

A recent example that the bottom line may not be the only determinant of executive success was reported in *The Wall Street Journal* (Brannigan & White, 1997). Ronald W. Allen, Chairman of Delta Air Lines, had turned the troubled aircarrier around after several years of heavy losses to earn record profits for eight straight quarters. He managed this through his "Leadership 7.5" program, whose central goal was to reduce operating costs by almost 30%, or $2 billion, to 7.5 cents per "available seat mile" (cost of flying one passenger seat 1 mile) from the previous 9.26 cents. Warnings that this goal was unrealistic were brushed aside, and Mr. Allen pushed forward plans to cut about 12,000 of Delta's 69,000 employees.

Experienced employees were dismissed and customer service scores began to plummet; on-time performance dropped precipitously, mishandled baggage complaints increased, and customers flew on dirty planes. Unions' organizers, who had never been a serious problem, came running.

Mr. Allen, faced with diving morale and soaring customer complaints, admitted that the drastic cost-cutting had upset employees, "but so be it." "So be it" buttons started showing up everywhere. Senior managers headed for the exits, tired of his caustic style, which included dressing down employees in front of others.

As problems mounted, Allen tried to reverse the decline in customer service and employee morale by ending staff cuts and postponing efforts to reach "7.5." He rolled out a package of wage increases, hoping to blunt the organizing drive. He thanked staff in memos and speeches for their sacrifices and talked of "rebuilding."

The employees, however, remained skeptical and scared. In a first-ever companywide employee survey, employees were asked to rate the effectiveness of the company's leadership. Almost half of all employees responded, and of those, 48% were unfavorable, 30% were neutral, and only 22% were favorable.

Leadership cannot be a popularity contest. Mr. Allen, however, cut into Delta's muscle, not just its fat, and his autocratic management style affected his ability to adequately lead. Delta's board of directors came to the same conclusion as they saw the company's superb customer service reputation tarnished, and its employees bitterly complain about declining morale. The authors of the newspaper article suggest that the board believed the turnaround had cost too much. Another possibility is that the board did not believe Allen could regain employee confidence, without which he could not effectively lead. Because of "an accumulation of abrasions over time" (p. A1), the board unanimously agreed that Mr. Allen's contract was not going to be renewed, and Delta retired him at age 55. The board stood up to the chairman not because Delta was going broke, but because he had broken the company's spirit.

The authors suggest that this is a surprising message in post-downsizing times. However, other businesses have learned the importance of customer satisfaction. The American auto industry continues to struggle against a perception that their product is inferior to imports and that customers are likely to be fleeced by fast-talking salesman. Krantz (1998) suggests that the importance of customer satisfaction will only grow in the future, particularly when there are few price differences between competitors. But employees cannot be expected to provide quality service when they are in a regressed and paranoid state.

Part of the responsibility must lie with management, which attempts to deal with complex problems through simple solutions. If the only option given serious consideration when budgets need to be trimmed is mass firings, management should not be surprised by the animosity and resentment their leadership engenders.

There seems to be an effort to separate senior management from staff and so shield them from the pain of those below. Not wanting to know may not be that unusual. Recently I heard about another regional hospital that was locally known for staffing problems and rapid turnover. They hired a consultant to question each nurse who had resigned over the previous 2 years because management now wanted to know why they left. Previously, no one had done exit interviews. Their reasoning: The hospital did not have enough manpower to find out why they were leaving. Clearly, this was an expensive undertaking to determine information that should have been a priority in the previous 2 years.

Employees place a high value on open communication and control over their job (Galinsky, Bond, & Friedman, 1993). The most frequently mentioned personal measure of success in this survey was the satisfaction of doing a good job. Making a good income was well down the list.

Krantz (1998) argues that when an intelligent work force is not expected to contribute, it increases employee dissatisfaction and is a gross mismanagement of a valuable resource. Had the VP consulted with the senior staff and been open to their comments, she would have learned that everything she was about to try had been tried just 3 years earlier by the last VP. Cut the patient-staff ratio and the RNs quit. LPNs will stay with the hospital through thick or thin because of their seniority (pension) and their limited options outside of the hospital. Unit managers end up burning out and quitting. Senior nurses refuse promotion to unit managers because they believe they will have to work 60 hours for 40 hours pay when there are staffing shortages. The VP denied this would happen when I suggested that this was the root of their resistance.

## CONCLUSION

Berg (1997), in his study of followership, points out that management too often views employees as sheep or lemmings, passive and obedient. Followers as a whole are demeaned by institutions and society, and managers split this role off from themselves. Employees need to be seen as cooperative, as team players, as learners and implementers, and as a valuable resource leaders can tap into during challenging times.

Steve Jobs, in a recent interview (Suarz, 1997) on the history of Silicon Valley, proposed that the traditional way management treats labor is no longer acceptable. Employees need to be treated with tremendous amounts of respect because they are the source of innovation; they are the intellectual capital. While Jobs was talking about computer programmers and engineers, the same is true for any work force made up of highly educated and trained professionals. To demean or ignore their input is to waste that capital. Nor is this view restricted to professionals. Don Soderquist, Vice Chairman of Wal-Mart, states: "We look at people as more than a pair of hands, we look to them as a source of new ideas" (Covey, 1997).

Without being valued for their contribution, employees will not invest in their work and increase the level of their performance (Krantz, 1998). Leaders need to get employees involved through true collaborative effort by reintegrating thinking *and* doing—a coherence of experience.

This "consultation" shows the legacy of an outdated leadership style that leads to staff regression, scapegoating, and a hateful, paranoid relationship toward authority. When a staff feels emotionally and intellectually disabled, their ability to collaborate across boundaries is restricted, and they are unable to link their work to a wider purpose. Feelings of powerlessness leave them unable to perform to the best of their abilities, which is a waste of valuable resources in a competitive market.

This new way of relating to employees will require a new type of leader. Krantz (1998), Miller (1998), Berg (1998), and Kram and McCollom Hampton (1998) all have similar constructions of such a leader: one who can be open, listen, and even be vulnerable. The popular press has a similar portrait; Nelson (1997) suggests the most important element for employees is to feel heard.

The leader who arrives with all the answers has no need for additional data. However, with constant change, learning becomes even more critical (Krantz, 1998). Employees need to be able to communicate without fear of reprisal, and the leader needs to tolerate the shame of not knowing, being vulnerable, learning from others, and being wrong in public. Allowing this more visible interior life and accepting the creativity in others will lead to more committed followers. Miller (1998) contends that this new leader must recognize the competence of each employee and be open to his or her own mistakes. Berg (1998) counsels leaders to accept their own weaknesses and shortcomings and to seek help when they are in trouble. No one has the ability to survive alone. By allowing others the opportunity to contribute their valued competence, the leader is the "good enough mother" (Winnicott, 1965). Leaders must nurture that which threatens to divide.

The employee has to feel safe in order to challenge the leader, which is likely only if a strong commitment already exists. When followers find their own voice and use it to express ideas, solutions, perspectives, and opinions, they take a courageous risk. This requires the leader to take the risk of following at times and not being concerned with how he or she looks to subordinates, peers, and superiors.

This is networking, not hierarchy. Massive information is available to be exploited if the leader can tolerate this challenge to authority. To some this may feel like rebellion. Managers often dread this bogus anarchy because of their own anxiety about the irresponsibility of those that work beneath them. What is proposed here is to promote autonomy and eschew dependence.

The current labor situation has included little wage pressure, even with low unemployment. This has been due, in part, to the high level of worker anxiety and job insecurity. Whether wage pressures return is uncertain. However, demographics suggest that competition for young, highly skilled, competent employees will only increase as the Baby Boomers age and are followed by significantly smaller cohorts. These employees are likely to be actively courted. If job satisfaction is the most important measure of personal success, the leader who can be open, listen, and be vulnerable will have a loyal work force of exceptional producers.

Loyalty, participation, and support are the bedrock of the leader-follower relationship. Employees need to feel heard and valued. When an organization faces hard times, communication needs to be opened in all directions. Creative solutions are possible; simple solutions for complex problems must be avoided. If all options are openly addressed and shared and cuts are still needed, the staff can then accept the inevitable with the belief that their leader did everything possible on their behalf. Better decisions, more willingly implemented, are possible after assessment and dialogue.

## REFERENCES

Berg, D. N. (1998). Resurrecting the muse: Followership in organizations. In E. B. Klein, F. Gabelnick, & P. Herr (Eds.), *Psychodynamics of leadership* (pp. 27–52). Madison, CT: Psychosocial Press.

Bion, W. R. (1961). *Experiences in groups.* London: Tavistock.

Brannigan, M., & White, J. B. (1997, May 30). "So be it": Why Delta Air Lines decided it was time for CEO to take off. *The Wall Street Journal,* p. A1.

Covey, S. (1997, August 29–31). How to succeed in today's workplace. *USA Weekend,* pp. 4–5.

Galinsky, E., Bond, J. T., & Friedman, D. E. (1993). *The changing workforce: Highlights of the national study.* New York: Families and Work Institute.

Klein, E. B., Gabelnick, F., & Herr, P. (Eds.). (1998). *Psychodynamics of leadership.* Madison, CT: Psychosocial Press.

Kram, K. E., & McCollom Hampton, M. (1998). When women lead: The visibility-vulnerability spiral. In E. B. Klein, F. Gabelnick, & P. Herr

(Eds.), *Psychodynamics of leadership* (pp. 193–218). Madison, CT: Psychosocial Press.

Krantz, J. (1998). Anxiety and the new order. In E. B. Klein, F. Gabelnick, & P. Herr (Eds.), *Psychodynamics of leadership* (pp. 77–107). Madison, CT: Psychosocial Press.

Miller, E. J. (1998). The leader with the vision: Is time running out? In E. B. Klein, F. Gabelnick, & P. Herr (Eds.), *Psychodynamics of leadership* (pp. 3–25). Madison, CT: Psychosocial Press.

Nelson, B. (1997). *1001 ways to energize employees.* New York: Workman.

Suarz, R. (Host). (1997, July 7). Silicon Valley's history. *All things considered.* Washington, DC: National Public Radio.

Winnicott, D. W. (1965). *Maturational processes and the facilitating environment.* London: Hogarth.

# 11

# AIDS and the Organization:

## A Consultant's View of the Coming Plague

*Burkard Sievers, Ph.D.*

### THE TIME OF AIDS

Organizations are built on a variety of differences, such as power, age, sex, knowledge, or sanity. In addition, such differences are often fictitious and are socially constructed. For example, in mental hospitals the differences between professionals and inmates often is an artificial one in regards to the madness these people have in common (Luske, 1990). As I have noted (Sievers, 1995a, 1995b), the obvious difference between nuns and inmates in a home for "fallen" girls may overshadow the fact that both establish partial relationships to men; in another case, an institution caring for foster families, it became evident that the helpers had much more in common with their clients than the apparent role differences would indicate.

Most fictitious, however, is the discrimination between "immortals" and "nonmortals" as it emerges as the predominant construction of contemporary organizations (Sievers, 1986, 1994). This is an outcome of the underlying splitting and the related idealization of people into heroes or devaluation of them as things. In enterprises, the illusion of the immortality of those at the top is derived from their identification with the immortality of the firm, giving others the fate of ephemerals, machines, or goods.

The artificial discrimination between immortals and mortals takes on a more literal meaning and further divides an organization when it is faced with the 21st century plague: Acquired Immune Deficiency Syndrome—AIDS. Sexuality, which for millennia has been mankind's predominant metaphor for vitality, pleasure, potency, creativity, and survival, now has received the connotation of a fatal disease that almost inescapably leads to death.

As AIDS is transmitted by blood, the latter undergoes a fundamental change of meaning, too. Whereas blood, both mythically and physically, has been a source of redemption (Jesus Christ), reliable friendship (blood brotherhood), invulnerability, satisfaction, sacrifice, or survival, it is now perceived as poisoned, polluted, or contaminated.

In the past, nakedness was an important concomitant of the excitement of sexual activities. At present it seems to have been replaced by the surgeon's glove and condom. Lust and spontaneity have thus become a matter of physical and rational control. There is truth in the saying that the world has changed since the discovery of AIDS more than a decade ago. Social consciousness of AIDS and an awareness that sexual relations have drastically changed are now apparent. "Now AIDS obliges people to think of sex as having, possibly, the direst consequences: suicide or murder" (Sontag, 1989, p. 72). For a short period, medicine, through contraception and the apparently easy curability of sexually transmitted diseases, "made it possible to regard sex as a adventure without consequences" (Sontag, 1989, p. 72). No more.

The fear of death—one's own in particular—may lead to the withdrawal from any communication and to concentration, like Narcissus, on one's own image. AIDS, as Sontag (1989) states, "further strengthens the culture of self-interest, which is much of what is usually praised as 'individualism.' Self-interest now receives additional endorsement as simple medical prudence" (p. 73). Whereas sex traditionally had been related to a whole variety of experiences, because of "safety" it is now reduced on a binary code basis: It is either safe or not (Herdt, 1992).

During epidemics in previous centuries *the disease* often became the incarnation of death. With AIDS it now seems as if *the Other*, the HIV-positive lover, coworker, neighbor, or friend, has become death's representative. Contrary to the classic iconography of the late Middle Ages in which the Black Death was the companion of 40 protagonists (Kaiser, 1983; Rosenfeld, 1974) who represented both Church and society, it is now the Other—as sexual partner, patient, colleague, blood donor, or family member—who as the potentially infected transmitter has become the carrier of death (Fineberg, 1988).

Unlike the time of plague, when the "Dance of Death" symbolized that escape was in vain, our infected "Other" is to be neutralized through avoidance or branding. Because of the dramatically long latency between infection and the onset of symptoms of AIDS, those who are identified as infected are treated as if they are ill. "Infected *means* ill, from that point forward" (Sontag, 1989, p. 32). And as soon as the illness becomes public, discrimination begins. "Those who test positive for it are regarded as people-with-AIDS, who just don't have it" (Sontag, 1989, p. 32). They must face all kinds of discrimination, dismissal from employment, and enormous economic consequences before they actually have to stop working (Crystal & Jackson, 1992).

But beyond mere avoidance and branding, "a good deal of mass media treatment of AIDS has constructed people with AIDS as 'other' than self" (Adam, 1992, p. 7). AIDS is seen, primarily, as a fate of people belonging to minority groups: prostitutes, homosexuals, drug addicts, and hemophiliacs. With the exception of the latter, who are regarded as victims, these infected minority groups are judged to be responsible for their own fate.

"Illness is said to be the outcome of their free choice of high-risk behavior" (Farmer & Kleinman, 1989, p. 146). In addition to the ascribed irresponsibility, this label implies a moral judgment, as a plague has often been seen as God's punishment for sin. This stigma of guilt and shame removes the uninfected's offering of sympathy and allows for guilt projections and self-exculpation.

Most recently, the medical discourse about AIDS has become the predominant one, and it has been increasingly tied to the martial metaphor. "The virus invades the body; the disease . . . is described as invading the whole society" (Sontag, 1989, p. 66). People dying from AIDS are regarded as if they had died on the battlefield in the fight against the disease. The 1992 memorial for AIDS victims indicated that 150,000 Americans had died—the equivalent of a war cemetery. The war and its dead concern distant multitudes, not the intimacy of the here and now. "The person with AIDS is annihilated as a subject and assigned the role of 'problems'" (Adam, 1992, p. 5). The various professions required to deal with AIDS "qualified the disease and operationalized the variables in our response to it; the person, *the sufferer,* and his or her intimate circle are left invisible" (Farmer & Kleinman, 1989, p. 151). If it is true that "we know ourselves only through the echo of the Other" (Cooper, 1983, p. 202), this echo is now disturbingly dissonant and remote.

Part of what makes AIDS different on a conscious and unconscious level in Western society is that it has undone 100 years of increasing life expectancy that instilled a new denial of our own mortality. We have since become used to the idea of dying late after retirement and have expatriated death out of our working lives (Sievers, 1994; Ziegler, 1982). In the time of vast cholera epidemics during the last century, where high mortality rates and short life spans were normal expectations, individuals were acutely aware of their own mortality. Now, people have again become aware that others die before the age of 30, and that they are not completely immune. "For generations of presuccessors it has been completely natural that death and dying could happen every time" (Imhof, 1991, p. 12). However, we have become accustomed in the last half century to thinking of ourselves as no longer subject to the incursions of such ills; death from acute infectious disease has seemed—like famine—limited to the developing world. Life-threatening infectious ills had become, almost by definition, amenable to therapeutic or prophylactic intervention. AIDS had reminded us that this sense of assurance might have been premature, the attitudinal product of a particular historical moment. AIDS has shown itself both a very traditional and a very modern sort of epidemic, evoking novel patterns of response and at the same time eliciting—and thus reminding us of some very old ones. (Rosenberg, 1989, p. 2)

In addition, contrary to a diagnosis of cancer, where the patient and significant others may derive some hope that the diagnosis may prove wrong or that some treatment exists that can cure or remit the cancer, the diagnosis of HIV, at least at this time, equals a death sentence whose execution is only postponed temporarily. But this in itself is problematical, as individuals may live 10 or 15 years after discovery of positive HIV status. "Because those who are symptomatic but not yet diagnosed with 'official' AIDS are in a particularly uncertain situation, it may well be the case that the dilemmas of uncertainty are even more severe and anxiety provoking" (Crystal & Jackson, 1992, p. 176).

The anxiety caused by an extended infectious period without AIDs symptoms may even lead to a reversal of values. The individual may feel that the justification for living is no longer sufficient, due to unemployment and lost relationships.

## AIDS AND THE ORGANIZATION

Although AIDS "has led to the formation of new organizational structures" (Ergas, 1989, p. 200), most of the research "has focused on individual behavior and change. Institutional analyses remain scarce" (Schneider & Huber, 1992, p. xxii). The following vignette will examine how AIDS can have a major impact on an organization and what is presumed to be "normal" and "fair."

The case material originates from an organizational role consultation during a working conference (Auer-Hunzinger & Sievers, 1991; Barry & Tate, 1988; Reed, 1976; Weigand & Sievers,

1985) with a psychologist working in an AIDS group home for HIV-infected individuals and those with AIDS. As in other role consultations, the thoughts developed here originated from our subjective views derived from experiences of the institutional matrix. Although the responsibility for these thoughts are mine, I am indebted to this particular Other. As working hypotheses, they are the outcome of action research.

The consultation began with an analysis of the different roles the psychologist had taken during the working conference. These roles were characterized by his joining a minority group during the institutional event which met outside the provided territorial boundaries; he also had been identified with the oldest member, a man in his late 50s who had been set up by the membership as a potential casualty. As this man took part in the role analysis group he enabled others to reintegrate their own madness previously projected into him and helped the psychologist to elaborate to what an extent he had identified his own childhood with the organizational drama in the conference.

The psychologist became aware how his father and his elder brother were the protagonists whom he had reactivated in the temporary system of the conference. He still was filled with rage toward his father, who had been absent most of his childhood. He had only entered into his life lately. At present his father was in a hospital recuperating from a stroke. The psychologist had intended to visit him at the end of the conference. His mother always had been the most important person in his youth, especially because his only brother was 12 years his senior. Although the psychologist was almost brought up as an only child, his life was affected by a brother so distant as to be beyond reach. In the relationship to the oldest conference member he had both projected the unattainability of these two male relatives and the sudden awareness of his father's mortality.

As we turned to his work situation in this charitable organization for HIV-infected individuals, it was evident that the obsession to be equal despite irreconcilable differences was a dynamic of how he had constructed his role in the institution. Working professionally with HIV-infected people he was, to a large extent, preoccupied with the experience of not being equal and therefore not accepted. So obsessed was he by the desire to bridge the gap that he longed to be mortally ill. We later realized the "perversity" of such a desire in regard to his working situation. To be mortally ill, without a doubt, means to face an early death.

The "primitivity" of this desire equals the wish of a psychoanalyst that Moeller (1992) richly describes. As this woman met privately with AIDS patients she felt an irresistible desire for unprotected intercourse with one of them. Although she was aware of the madness of this act, she nevertheless felt a desperate longing.

These two episodes illustrate the difficulty of professionals in managing their own dynamics in order to work in a role, derived from the institution's primary task, of helping people to keep on living and to die with dignity. Despite the "primitive" desire for fusion with the "terminal outcasts," the management of one's working role requires the acknowledgment that the clients are different.

In the role analysis group, we explored the psychosocial dynamics of the HIV group home and realized that the management of differences required superhuman competence and energy. What made the work more difficult was that in addition to the people in the institution, the sponsoring charity consisted of both infected and noninfected members. Unlike most AIDS organizations, in which the clients, but not the helpers, are infected, this institution was founded by a group of both HIV-positive and HIV-negative individuals with no explicit role differentiation. This institution was both a self-help group of infected individuals and a benevolent organization in which noninfected professionals provided help for the infected. This meant that the board and the staff included people who knew who was infected and who was not. Although this constellation during the early years seemed to have some advantage as far as people's commitment, it soon caused havoc.

As time passed, the HIV-infected people who previously had not been ill either were suffering AIDS symptoms or had died. This led to changing roles, from being managers to being sick and dying patients. The institution felt obliged to maintain the employment of the seriously ill, while at the same time it had to reduce the help it intended to provide. The noninfected staff had to accept part-time employment, as the sick could not be dismissed. In effect, because the work still needed to be done, the noninfected were expected to contribute part of their time and work without compensation. This created a vicious cycle. Most of the noninfected individuals were supposed to earn their living through this work, and as they were already earning less than in other comparable jobs, they felt neglected. One can only imagine how the infected management staff, who were employees as well and had themselves helped found the charity, dealt with the increasingly difficult economic situation and the needs and demands of those who were noninfected.

Already used to the experience of not getting any gratitude from the HIV-infected, the noninfected staff now had to cope with envy. They had been aware of being envied due to their longer life expectancy; now they felt envy themselves about their former colleagues, who, despite their inability to work, received a higher income, letting them, the uninfected, cope with the burden of underemployment. Though professional AIDS caregivers are always concerned with not having enough time for the particular personal situation of their patients (Waldvogel & Seidl, 1992), the increasing demands, in this case, led the caregivers to focus on some patients while ignoring others. The noninfected caregivers could only appear as inadequate. Further, they had to face the reproach that they selfishly used the fate of the infected for their own gain.

Although the dynamics of this charity were similar to those of voluntary institutions (Chatwin, 1991), it differed, however, from the majority because of the awareness of mortality. Regardless of the more formal differences between staff, the most important fact remained whether a particular person was infected and therefore faced a more imminent death. Whether one was "positive" or "negative" also tainted a person with moral connotations; however, it was the reverse of society's judgment of good or bad. Having to face an early death "normally" is regarded as negative. The HIV-infected regarded themselves as superior, thus making the noninfected accept the definition of themselves as being inferior. To identity oneself as positive had become a defense against the fatal illness.

The wish to identify oneself as positive caused a competition and mutual devaluation. During the early years of this institution, infected people had used their positive test result as a source of energy to provide help for those who were further progressed in suffering from the disease; now the infected created a dependency culture (Bion, 1959). This was particularly true of the drug users, who would discriminate against the infected homosexual men. They viewed themselves as victims and idealized the consumption of drugs as an expression of potency and protest against society. They sustained this image to confirm that they had nothing in common with gays but a virus. Regarding themselves as innocent victims, the positives demanded services that the negatives were supposed to provide for them. The more the original idea of a voluntary organization fell into obscurity, the more the caregivers were regarded as representatives of society's hostility toward people with AIDS.

It appeared that the splitting among positives and negatives served as a defense against the awareness that the infected had to face an earlier death than the noninfected. Although many people had died from AIDS, at least to the professional caregivers it appeared that the organization was unable to cope with death. Death notices were not made public and mourning ceremonies were unknown. Every death was to be dealt with individually. Because the positives were preoccupied in either acknowledging, or avoiding, the thought of their own death, the losses had to be managed individually by the negatives. The negatives had to contain all the emotions, and not express them. To the extent that being infected by HIV was regarded as superior, there was no space for the negatives to express their own anxieties about the risk of infection through contact with positives. Of course, repression of such powerful feelings led to depression among the HIV-negative staff. Further, for the noninfected staff, their daily experience included such a high amount of strain, including the work's constant focus on death anxiety, that the thought of escaping appeared as the final attempt to maintain their autonomy.

Labeling of the noninfected as inferior also made it difficult for the professional caregivers to acknowledge their own competencies, strength, and potency. The obvious difference had to be hidden: The infected were entering into an irreversible phase of dying while the uninfected would live on. To the extent that these different awarenesses were ignored, the difference itself could not be used creatively as a source of energy.

This ignorance serves to maintain a psychic prison (Morgan, 1986) in which all have to face imprisonment, with the difference being that the many have death, and the few, life sentences. According to the dynamic of this split and the accompanying projective identifications, it is the latter who are not supposed to experience despair in their search for meaning, and so become the object of envy for the former. The noninfected have a chance to get rid of AIDs by leaving the institution, whereas others are stuck in the seemingly futile hope that they will be the first ones to survive the disease. The more the difference in the life spans of the infected and the noninfected are ignored, the higher the probability that the HIV-negative caregivers would end up working below their potential.

There is no doubt that the daily work in this institution always had priority. Suffering individuals were given the attention they needed and administrative matters concerning the charity as a whole were frequently postponed. However, the management in this case appeared to be preoccupied with the mortality of its members and to neglect the institution's survival. Under normal circumstances a mature management symbolizes reliability and containment for the dependency needs, anxieties, and projections of members. In this circumstance, this capacity was questioned. Management was either invisible to, or despised by, staff and the infected. The situation was further exacerbated because the future existence of the institution was permanently at risk.

The management of this organization had no means to extend the lives of the infected, and was thus likely to appear incompetent or selfish. To the infected, management appeared greedy, unable to satisfy their demands. Management was seen as failing them. To compensate for their failure, management coped with their guilt through working extended hours.

Contrary to individual death, which is inevitable, the potential death of the institution was not preordained, thus nurturing the hope that death could be prevented through further activities. Management's perceived failure to protect the institution was, in effect, a failure to nurture the hope that death could be averted. To the extent that management worked in the survival of the institution, it could, and would, also disengage from the suffering of the infected. To the extent that a few were mainly concerned about institutional survival, the basic equality that all are mortal was further neglected. If death of the institution preoccupies management, a catastrophe that cannot be prevented, management negates the reality that individual death is a fact of life.

Management failed those suffering from AIDS and failed the staff that served them as well. For the management in this institution, it was the very competence to manage oneself in role that was at risk (Lawrence, 1979). The potential death of the institution thus became a new trauma for all involved, a cumulative trauma (Khan, 1974). The experience of the potential shutdown not only merged with the trauma of the expectable death of the infected, but also with the reactivation of old traumatizations from previous losses (Segal, 1972; Weinel, 1992; Winnicott, 1974). Unlike the management of a traditional health care institution, the management of this organization was itself representative of illness and death. The anxiety about the institution's survival was representative of the anxiety about one's own death.

## CONCLUSION

This case illustrates how an institution dealing with AIDS is changed and challenged as an organization. This institution had become pathological, as Steiner (1988) describes it. In a pathological organization the paranoid-schizoid and depressive positions are in equilibrium. The pathological organization functions as a defense against the fragmentation of the former and the anxiety of the latter position. Insofar as pathological institutions lose the capacity to acknowledge anxieties as real and to contain them, they are permanently in jeopardy of losing any parameter for a "reality principle" or for maturity. The dynamic of these institutions is also the same as that of a psychoanalytic group in which

> at every point in time ... the words, ... emotions, ... and thoughts of group members circulate around one central unconscious theme.... Each one processes the dominant theme according to his own biography because the anxieties and desires underlying the defenses are highly individual. This may lead to the phenomenon that all are expressing the same basic theme although, on occasions, for an external they appear to talk about totally different issues. (Moeller, 1992, p. 5)

AIDS in the organizational context thus contributes to creating a "curious community of each one against the other, a security of suspicion, a solidarity of paranoia.... Distance and intimacy, mistrust and the feeling of belonging together mark the hidden quality of relations.... AIDS is creating a kind of magnetic field among us through which we unconsciously are controlled" (Moeller, 1992, p. 5).

Management of responsibility became difficult in this organization. Responsibility in the context of AIDS (Bayer, 1989) is a matter of life and death. It is particularly difficult for those who are drug users or hemophiliacs to accept. They resist responsibility for their illness because they are victims. It is true that "drug use, and the spread of HIV disease to the sexual partners of drug users occur in zones of urban poverty, poor health care, social disintegration, and the epidemic spread of other diseases, including diseases that interact with HIV and facilitate its spread" (Lindenbaum, 1992, p. 325). Because it is difficult for the management to substitute for the missing responsibility of these patients, the chances are high that the lack of responsibility then is projected into the noninfected staff and the government. In face of the fact that the financial resources for people affected by AIDS are scarce almost everywhere, the reproach toward society for its irresponsibility creates further guilt feelings, which, as they cannot be accepted, then are converted into further neglect.

The threat of AIDS is that in the long term it destroys the ability of the body to defend against contamination from the environment. It also seems that this organization is characterized by an irreversible weakness in maintaining its "immunity." It further seems that the noninfected people working there have in their daily experience such a high amount of strain (Fineberg, 1988) and death anxiety that the thought of escaping appears as the final attempt to maintain their autonomy. What Witkin and Poupart (1986) have described for an abortion clinic—that the members of staff, because of extremely high emotional involvement, can only be effective for a limited time—may be more so in any AIDS institution.

Most uniquely, this case demonstrates that the usual split between immortals and mortals becomes drastically significant if part of an organization's membership has to face an early death. Contrary to Greek mythology and its underlying dichotomy of immortals and ephemerals, we have the reminder that all humans inevitably are mortal. This recognition seems to be more important because AIDS not only contributes to an increased societal thematization of mortality, but also serves as a defense insofar as it nurtures the illusion that by only avoiding this disease one may live forever. "The ultimate pain of the epidemic is inflicted on those who are living with and dying of HIV/AIDS; other pains, even those experienced by those most intimate with those who are living and dying with the disease, usually recede, albeit slowly, with forgetfulness" (Gagnon, 1992, p. 38).

AIDS has changed the consciousness of people and has had an enormous impact on those who are affected by the illness. What has become obvious is that "the AIDS epidemic is forcing us to change the way we think about and study culture" (Herdt, 1992, p. vii). At the same time, the attempt not to let AIDS become a global apocalyptic threat (Sontag, 1989) forces on us a cultural transformation that by far goes beyond the mere changing of individual patterns of sexual behavior. Especially in the United States, the "emergence of a culture of responsibility and restraint" (Bayer, 1989, p. 80) will only have a limited impact if it is not accompanied by socioeconomic changes that will decrease intravenous drug use and teenage pregnancy. What has become evident is that we experience a significant transformation from private to social knowledge. As public health campaigns attempt to educate a majority of a population and have no means to differentiate according to age, even young children are introduced into the "mechanics" of sex and drug use. This leads to the consequence that children often "know" more about sex than their own parents, but that their future sexual activity is less driven by passion, desire, and love than by the awareness of its risk—and the anxiety that the Other, more than an "object" of love, may be contaminated and a transmitter of the virus. The risk displaces what previously has been regarded as the joy of sex.

The appearance of the AIDS epidemic has an impact on the emergence of a variety of new organizations and the transformation of existing ones. It characterizes the "Time of AIDS" (Herdt & Lindenbaum, 1992) that the "Shadow of AIDS" (Lindenbaum, 1992) has not only fallen on individual people with AIDS and those close to them, but also touches organizations, and merges with the organization shadow and its irrational forces (Bowles, 1991).

The particular impact AIDS has on organizations and their cultures apparently varies to the extent that the disease is part of an organization's primary task. The preceding example of the AIDS institution indicates how an organization's culture is affected if the care for infected people is the primary task. In traditional systems of health care, education, or prisons, the mere possibility that a client may be infected often creates such disturbance that the attempt of excluding further transmission leads to the creation of structures that are intended to defend against the anxieties (Menzies, 1970) that effectively prevent further infection. The cultural impact of AIDS on other organizations, which do not have as their primary task the treatment of diseases, can only be guessed at the present moment. These organizations will be affected in the same way as social life in general, that is, organizational life is determined by the anxieties activated through the potentially infected. Only in special industries, like arts, sports, or entertainment, does the fact that a prominent figure has become ill because of AIDS get attention, whereas in other cases it is hidden.

The appearance of AIDS is proof that mankind has not been

liberated from its biological constitution (Rosenberg, 1989); "epidemic diseases . . . are no more relic(s) of the pre-industrial age" (Evans, 1988, p. 124). Whether this experience will be regarded as a challenge or as humiliation, and thus will be displaced, cannot be foreseen. Farmer and Kleinman (1989) are skeptical whether we and our organizations will be able to react any differently to "what people with AIDS are telling us about their experiences" as to maintain our "now dominant definitions of reality and business-as-usual responses to human misery" (p. 158). Our inability to respond differently will for the generations that follow become "a telling indicator of the constraints on the human spirit in our times" (p. 158).

It is difficult to be optimistic. The metaphor of a "war against AIDS" has previously been mentioned. While not discounting the monumental suffering and death caused by AIDS to date and that will be caused in the near future, war, itself, continues to confront us on a day-by-day basis with countless deaths. Our ongoing individual and collective helplessness in regard to the wars in the former Yugoslavia, Somalia, central Africa, Cambodia, and elsewhere encourages our indifference and the creation of a fictitious reality in which we are immortal. While our denial and rationalizations may provide some small amount of comfort and insulation, it is clear that while all people are mortal, some are more so.

The psychologist with whom I collaborated left the institution and is now working in another field. The objective reason was that his job was no longer funded, although he seemed relieved to have this reason.

## REFERENCES

Adam, B. (1992). Sociology and people living with AIDS. In J. Huber & B. E. Schneider (Eds.), *The social context of AIDS* (pp. 3–18). Newbury Park, NY: Sage.

Auer-Hunzinger, V., & Sievers, B. (1991). Organisatorische Rollenanalyse und-beratung. Ein Beitrag zur Aktionsforschung [Organizational role analysis and consultation. A contribution to action research]. *Gruppendynamik, 22,* 33–46.

Barry, T., & Tate, D. (1988). Success in a new task. A role consultation. *Management Education and Development, 19,* 215–226.

Bayer, R. (1989). AIDS, privacy, and responsibility. *Daedalus, 118,* 79–99.

Bion, W. R. (1959). *Experiences in groups.* London: Tavistock.

Bowles, M. L. (1991). The organization shadow. *Organization Studies, 12,* 387–404.

Chatwin, M. E. (1991). *Creating havoc: Coping with voluntary organizations emotional states.* Paper presented at the Ninth SCOS International Conference, Copenhagen, Denmark.

Cooper, R. (1983). The Other: A model of human structuring. In G. Morgan (Ed.), *Beyond method: Strategies for social research* (pp. 202–218). Beverly Hills, CA: Sage.

Crystal, S., & Jackson, M. (1992). Psychosocial adaption and economic circumstances of persons with AIDS and AIDS-related complex. *Family and Community Health, 12,* 77–88.

Ergas, Y. (1989). The social consequences of the AIDS epidemic: A challenge for the social sciences. In M. W. Rile, M. G. Orey, & D. Zablotsky (Eds.), *AIDS in an aging society: What we need to know* (pp. 127–146). New York: Springer.

Evans, R. J. (1988). Epidemics and revolutions: Cholera in nineteenth-century Europe. *Past and Present, 120,* 123–146.

Farmer, P., & Kleinman, A. (1989). AIDS as human suffering. *Daedalus, 118,* 135–160.

Fineberg, H. V. (1988, October). The social dimensions of AIDS. *Scientific American, 269,* 106–112.

Gagnon, J. H. (1992). Epidemics and researchers: AIDS and the practice of social studies. In G. Herdt & S. Lindenbaum (Eds.), *The time of AIDS: Social analysis, theory, and method* (pp. 27–40). Newbury Park, NY: Sage.

Herdt, G. (1992). Preface. In G. Herdt & S. Lindenbaum (Eds.), *The time of AIDS: Social analysis, theory, and method* (pp. 1–7). Newbury Park, NY: Sage.

Herdt, G., & Lindenbaum, S. (Eds.). (1992). *The time of AIDS. Social analysis, theory, and method.* Newbury Park, NY: Sage.

Imhof, A. E. (1991). *Ars Moriendi: Die Kunst des Sterbens einst und heute* [Ars Moriendi: The art of dying once and today]. Vienna: Bohlau.

Kaiser, G. (Ed.). (1983). *Der Tanzende Tod. Mittelalterliche Totentänze* [The dancing death. Medieval death dances]. Frankfurt/M.: Insel.

Khan, M. M. R. (1974). The concept of cumulative trauma. In *The privacy of the self: Papers on psychoanalytic theory and technique* (pp. 42–58). New York: International Universities Press.

Lawrence, W. G. (1979). A concept for today: The management of oneself in role. In W. G. Lawrence (Ed.), *Exploring individual and organizational boundaries: A Tavistock open systems approach* (pp. 235–249). Chichester, UK: Wiley.

Lindenbaum, S. (1992). Knowledge and action in the shadow of AIDS. In G. Herdt & S. Lindenbaum (Eds.), *The time of AIDS: Social analysis, theory, and method* (pp. 319–334). Newbury Park, NY: Sage.

Luske, B. (1990). *Mirrors of madness: Patrolling the psychic border.* New York: Aldine de Gruyter.

Menzies, I. E. P. (1970). *The functioning of social systems as a defence against anxiety. A report on a study of the nursing service of a general hospital.* London: Tavistock Institute of Human Relations.

Moeller, M. L. (1992). Der Tod und der Trieb: Die Betreuung von AIDS-Kranken zwischen Professionalität und persönlichem Engagement [Death and desire: The care for AIDS patients between professionality and personal involvement]. In M. Ermann & B. Waldvogel (Eds.), *HIV-Betroffene und ihr Umfeld. Ergebnisse aus psychosozialer Forschung und Praxis* (pp. 1–17). Berlin: Springer.

Morgan, G. (1986). *Images of organization.* Beverly Hills, CA: Sage.

Reed, B. (1976). Organizational role analysis. In C. L. Cooper (Ed.), *Developing social skills in managers* (pp. 89–102). London: Macmillan.

Rosenberg, C. E. (1989). What is an epidemic? AIDS in historical perspective. *Daedalus, 118,* 1–17.

Rosenfeld, H. (1974). *Der mittelalterliche Totentanz: Entstehung, Entwicklung, Bedeutung* [The medieval dance of death: Origin, development, and meaning]. Cologne: Bohlau.

Schneider, B. E., & Huber, J. (1992). Introduction. In J. Huber & B. E. Schneider (Eds.), *The social context of AIDS* (pp. xiii-xxiii). Newbury Park, NY: Sage.

Segal, H. (1972). A delusional system as a defense against the re-emergence of a catastrophic situation. *International Journal of Psycho-Analysis, 53,* 393–401.

Sievers, B. (1986). Beyond the surrogate of motivation. *Organization Studies, 7,* 335–351.

Sievers, B. (1994). *Work, death, and life itself: Essays in management and organization.* Berlin: de Gruyter.

Sievers, B. (1995a). Characters in search of a theater: Organization as theater for the drama of childhood and the drama at work. *Free Associations, 5*(34, Pt. 2):196–220.

Sievers, B. (1995b). Organizational culture and its discontents. In T. C. Pauchant (Ed.), *In search of meaning: Managing for the health of our organizations, our communities, and the natural world* (pp. 271–292). San Francisco: Jossey-Bass.

Sontag, S. (1989). *AIDS and its metaphors*. New York: Farrar, Straus & Giroux.

Steiner, J. (1988). The interplay between pathological organizations and the paranoid-schizoid and depressive positions. In E. B. Spillius (Ed.), *Melanie Klein today: Developments in theory and practice. Vol. 1: Mainly theory* (pp. 324–342). London: Routledge.

Waldvogel, B., & Seidl, O. (1992). Belastungen und professionelle Bewältigungsformen von Ärzten und Kranken-pflegekräften, die Patienten mit AIDS betreuen [Burden and professional ways of coping among physicians and nurses caring for patients with AIDS]. In M. Ermann & B. Waldvogel (Eds.), *HIV-Betroffene und ihr Umfeld. Ergebnisse aus psychosozialer Forschung und Praxis* (pp. 33–45). Berlin: Springer.

Weigand, W., & Sievers, B. (1985). Rolle und Beratung in Organisationen [Role and consultation in organizations]. *Supervision, 7,* 41–61.

Weinel, E. (1992). HIV-Positive in psychoanalytisch orientierter Langzeittherapie [HIV-positives in psychoanalytically oriented long-term therapy]. In M. Ermann & B. Waldvogel (Eds.), *HIV-Betroffene und ihr Umfeld. Ergebnisse aus psychosozialer Forschung und Praxis* (pp. 137–142). Berlin: Springer.

Winnicott, D. W. (1974). Fear of breakdown. *International Review of Psycho-Analysis, 1,* 103–107.

Witkin, A., & Poupart, R. (1986). Shadows of a culture in "native" reflections on work in an abortion clinic. *Dragon, 2,* 56–72.

Ziegler, J. (1982). *Die Lebenden und der Tod* [The living and death]. Frankfurt/M.: Ullstein.

# 12

# Training Group Therapists for the 21st Century

*Mary Nicholas, Ph.D., and
Robert H. Klein, Ph.D.*

### GROUP THERAPY: WAVE OF THE FUTURE

To the satisfaction of the few psychotherapists who have practiced, taught, and written about group therapy over the years, it seems this powerful psychotherapeutic modality has finally come into its own. For decades in the mental health fields (with the exception of chemical dependency settings) group therapy has been considered an adjunct, if not inferior, form of treatment. It is only in the past 10 years, however, that group therapy has become paramount in the public sector, and it is growing widely in the private practice arena as well.

While the upsurge in group therapy represents, in our view, a change in the right direction, the motivations behind it are less philosophical than economic in nature (Klein, 1993). Group therapy's rapid growth and robust future are based primarily on the fact that it is considerably cheaper and more efficient than individual or even family therapy. Putting seven or eight people together with one or two therapists for an hour is more lucrative for the provider and less expensive for the patient and his or her insurance company than one patient per therapist per hour. While programs based on the model of therapeutic community—notably chemical dependency and partial hospital programs—enthusiastically endorse group approaches, most mental health systems still formulate their programs based on the medical model, which is, conceptually at least, somewhat antithetical to the nonhierarchical, patient-helping-patient approach represented by group therapy. Additionally, the majority of therapists, having been trained in individual models, still believe that the benefits of group therapy such as support and feedback are outweighed by the reduction and dilution of individual attention that each patient receives from the therapist in the group setting (Alonso, 1993).

The fact that the shift in favor of group treatment is pragmatic rather than paradigmatic and is based on economic rather than clinical exigencies raises some interesting challenges for the training of psychotherapists, both now and for the 21st century. First of all, while the number of clinicians doing group therapy around the country is already huge and still growing, few of them have received training. Second, there is a lack of consensus about exactly what constitutes group therapy; in fact, group therapy is not one entity, but, rather, encompasses a wide variety of treatments—many of them differing widely from one another in focus, theoretical basis, and technique. In this chapter we will sort out some of the complexities of group treatment as it exists today and articulate what we think is needed and will be needed in the way of training in the next century.

## THE SCARCITY OF TRAINED GROUP THERAPISTS

The pervasive ethos regarding group therapy before so many clinicians were called upon to actually do it was that it was really fairly simple. The myth that still shrouds group therapy in much of the mental health world is that group therapy just "happens" when several patients are put together in the same room with a therapist. Many therapists are still surprised to learn that there is a body of theory and an organized system of skills that clinicians need to learn in order to become group therapists. The predominant pedagogy of group therapy in the mental health world today is what we call the "nothing-to-it" theory. Since there is nothing to "it" (doing group therapy), why read about it or get trained and supervised? It is hardly surprising that managed care and insurance companies are ill informed as to the value of group therapy, when so many clinicians themselves seem unable to describe it cogently, much less practice it skillfully.

Until recently a trained group therapist was a rarity. Group therapy training is infrequently offered and almost never required in graduate programs for mental health professionals (some advanced psychiatric nursing and PsyD programs are the exception in this regard). Most therapists are still learning group therapy on the job, in their graduate school practicum, or through internship experiences. The problem is intensified since there are very few qualified group therapy supervisors in mental health settings. Many, if not most, clinicians are learning group therapy inadequately from people who were never trained themselves. As trainers we find that we are often called upon to teach "old dogs new tricks," a process that is not always smooth.

## Obstacles to Closing the Training Gap

In trying to teach the theory and practice of group therapy to therapists who have already been running groups "by the seat of their pants," we have run into some interesting problems that have affected the way we teach.

We have found that the untrained group practitioner or supervisor may or may not know his or her limitations and may or may not feel comfortable acknowledging that he or she could benefit from more training. As educators we have learned to be sensitive to the totally understandable need of veteran clinicians not to feel deskilled by incoming "experts." We have had to recognize that many line-staff therapists, while interested in the practicalities of group therapy, are intimidated by, and tend to avoid, theory. In an extensive 9-month training program that we conducted at a large mental health facility, the didactic sections received a tepid response; no one seemed interested in doing any reading and not one person took a note during a lecture. On a more positive note, however, while most are not particularly intellectually motivated, many of the clinicians we train are eager to improve their skills and are highly receptive to what we have to offer in that regard. Many individuals and institutions who would not have considered doing so 5 years ago are seeking our group therapy training and consultation services.

A second major problem we have encountered is that incompetence in group therapy has become (forgive the pun) institutionalized. Mental health environments tend to maintain shockingly low and/or flimsy standards of competence in group therapy. While an unlicensed therapist would never be permitted to do individual or family therapy, just about anyone is considered qualified to do group therapy. A student in a group psychotherapy course for PsyD students offered by one of us (MN) at the University of Hartford was asked by his supervisor at his practicum site to take over the men's group from someone named Joe, who apparently had been leading the group for several months. Joe, the intern learned, was the agency's receptionist. In another instance a therapy patient of MN's reported going to a local outpatient clinic to enroll in what was advertised in the local newspaper as a "therapy group for depressed women." The group, it turned out, was leaderless—organized by the agency and left to run itself, with one of the depressed members designated responsible for setting up coffee.

Many departments of psychiatry at major teaching hospitals are equally negligent regarding quality of care when it comes to group therapy. As clinical faculty in the Department of Psychiatry at Yale, both of us are regularly asked to supervise 3rd-year psychiatric residents in the outpatient department who have not had a single hour of group therapy training prior to starting their groups, which they do several weeks before meeting us. One of us (MN) was recently asked to do supervision via live observation of two 2nd-year residents who were the designated "team leaders" in a prestigious university-based research and treatment inpatient unit. The residents had received no training in individual psychotherapy, much less in group therapy, prior to becoming team leaders. MN decided it would be more useful to co-lead the group with the doctors, so that they could have some exposure to how group therapy is supposed to work. This offer was refused by the Department of Psychiatry on the theory that it would undermine the credibility of the residents on the unit. (As if the patients could not tell that their group therapists did not know what they were doing!) MN told the agency that it would be unethical of her to pretend to supervise untrained clinicians, but instead offered to put together training and orientation resources for new group therapists at the facility.

Needless to say, we have realized that our role as trainers involves educating the systems in which we teach about the value of group therapy and the complexities involved in its practice.

## Therapeutic Consequences of the Scarcity of Trained Group Therapists

Yalom (1995) has shown that properly run therapy groups have many therapeutic benefits for patients, including a sense of universality, instillation of hope and awareness of how one affects and is affected by others, and a chance to accept and reintegrate previously disowned parts of the self. We strongly believe that the therapeutic factors are consistently mobilized only in groups run by clinicians who can understand and manage the powerful dynamics of group process in the interests of members' treatment. It is no simple matter for therapists to learn how to work in the here and now and to recognize and selectively utilize phases of group development, dynamics, and possibilities for change on the individual, interpersonal, and group-as-a-whole levels. Unfortunately, very few practitioners have ever been members of a group, so they have no idea what patients are experiencing and no sense of the powerful potential group therapy has for helping clients. Having never attended sensitivity groups or Tavistock training, the naive group therapist overvalues the amorphous antidote known as intermember support, and shies away from anything that feels like negativity or conflict. Yet suppressed anger increases the stress of group members and erodes cohesiveness, often leading to iatrogenic effects on members, whereas the successful resolution of conflict enhances cohesiveness, thereby building the therapeutic effectiveness of the treatment for its members. Many leaders of short-term groups unwittingly prolong Bion's (1961) "basic assumption dependency" on the leader and unconsciously encourage "flight" or avoidance phenomena for the few weeks allotted the group by maintaining an exclusively psychoeducational focus, that is, providing many tasks and exercises, and keeping interpersonal interaction to a minimum. In our opinion, such a restrictive approach misses the whole point of group treatment.

Sometimes, by chance, the dynamics of groups led by untrained therapists have a powerful and positive effect on members—the mechanisms of constructive feedback and conconfrontation take place spontaneously or the group takes a person to a deeper level of experience—but, when this happens, the uneducated therapist does not know how or why it occurred and, therefore, cannot replicate these curative processes. Often groups run by untrained therapists fall flat either because the clinician inserts himself or herself into the process when the group could do better on its own and/or because the therapist persistently talks to one individual at a time instead of facilitating interaction among all the members.

We need to identify which skills and theoretical concepts are generic and vital for group therapists. Before we do this, however, we need to define more clearly what we mean by group therapy.

## THE MANY FACES OF GROUP THERAPY

### Five Key Dimensions

Complicating the task of trying to properly equip the clinicians of today and tomorrow with good group therapy skills is the fact that there is little agreement in the mental health field as to what group therapy *is*. The activities that pass as group therapy in the 1990s vary widely, and there is tremendous confusion about which, if any, are truly valuable and why (Dies, 1992). Differences of opinion about group therapy's value in the mental health marketplace is reflected in the fact that some managed care or insurance companies reimburse it on a par with individual therapy and others refuse to reimburse it at all. All types of treatments are lumped by insurance companies under the category of group therapy regardless of what service they provide or whom they serve.

Further complicating the situation is the lack of differentiation in the mental health field between the (1) *modality* being used, (2) *problem* the group is supposed to address, (3) *composition* of the group, and (4) *setting* in which the group takes place. A clinician might be asked to lead an eating disorders group without reference to what sort of group therapy method was being used (e.g., cognitive-behavioral, interactional), as if all types of treatment for eating disorders patients were the same. In this case, also, it is unclear whether the group is supposed to treat the person or the behavior. Is the therapist treating the eating disorder as part of a larger syndrome, or is the clinician teaching the members certain behavioral changes to make in order to eradicate the eating disorder? A plethora of legitimate and not so legitimate forms of group therapy are being applied pretty much willy-nilly to treatments groups organized around a large variety of problems and situations. Groups are formed to deal with particular psychopathologies (depression, eating disorders, addictions); traumas (rape, childhood sexual abuse), developmental challenges (widowhood, adolescence, divorce, parenting, grieving); or serious physical illnesses (AIDS, breast cancer); and no one, including the therapist, is clear about what group approach is being taken and why. An important dimension within problem or psychopathology is level of chronicity, which has implications for style of leadership and group duration (Piper and McCallum, 1994). Further complicating the picture is that virtually any combination of group composition, problem, and modality might be found in each type of mental health and chemical dependency setting: outpatient, partial hospital, or inpatient.

A fifth variable that interacts with all four of the others—modality, composition, problem, and setting—is *time*. In all sectors of psychotherapy practice, third-party payers are dictating the length of treatment time to practitioners. In inpatient settings, due to increasingly shorter lengths of stay, group therapists may see patients only two or three times during their hospital stay, and a group may have little or no continuity of population from one meeting to the next. In outpatient settings, therapists may be asked to provide justification for continued treatment after every two or three sessions, or be refused payment for a group beyond five meetings. All these time factors contribute to changes in group composition and size that are often outside of the therapist's control. The clinician is thus required to make each session cohesive and meaningful on its own, which necessitates the use of directive techniques and warm-ups that in longer term groups are generally bypassed in favor of a more patient-centered nondirective leadership.

Clearly, training group therapists in the 21st century will involve arming them with models that not only are brief but virtually instantaneous.

## Theory-based and Nontheory-based Group Therapies

In addition to five dimensions of therapy groups, we make a distinction between theory-based group therapies and nontheory-based group therapies, and believe that training should focus on the former.

Theory-based group therapy, in our view, is any group treatment that is based on a respected psychological theory and that is conducted by a trained therapist who is schooled in that model of treatment. The key difference between theory-based group therapies and the *ersatz* models is that the former provide opportunities and challenges for therapeutic change that do not exist automatically in any conversation and group of people. By contrast, nontheory-based group therapies are just like any social group except that a therapist is present who, because he or she is being paid to be there and is not sure why, will try to do individual treatment within the group, and/or get people to talk to each other without any particular purpose in mind.

Theory-based group therapy has as its goal some defined therapeutic change for patients (and this may differ with different models) that goes well beyond mere support. A feeling of not being alone with one's problems, which seems to be the generic meaning of support when people are talking about support therapy groups, is necessary but not sufficient for effective psychotherapy; therefore, a treatment that limits its purpose to providing support is not a theory-based therapy group. While some theory-based group therapy may be very supportive, practices in such groups are informed by a body of theory and geared to addressing carefully identified target symptoms and maladaptive modes of coping (Klein, 1979). It is important that the mental health community and public be disabused of the notion that therapy groups simply provide support.

Although every effective therapy group must establish a stable holding environment of trust, safety, cohesiveness, and freedom of expression within which the group can do its work (Klein, Bernard, & Singer, 1992), a truly therapeutic group is not always supportive, but rather confronts members with their rigidities, blind spots, and hypocrisies. To work properly a group must sometimes be challenging and difficult for its members. The reward the patient receives from group therapy derives from the lessons learned from working through the confrontations that inevitably occur when people are encouraged to be open and honest with each other, and these incidents can be exceedingly jarring and painful. The connectedness, or universality that patients report that they find in group therapy (Yalom, 1995) can only occur if the group interaction is highly authentic (Ormont, 1993). Group cohesiveness is the only variable that correlates with therapeutic effectiveness for its members (Budman et al., 1989; Yalom, 1995), and is the product of honesty, self-disclosure, therapeutic focus, trust, and caring. None of this process occurs automatically, but must be diligently worked for by therapists and members alike, and this involves facing and talking about hostility, lack of authenticity, and mistrust when they occur. The easy kind of support that comes from being allowed to blame other people or fate for one's problems is corrosive, not therapeutic.[1]

A type of group that has been passing as psychotherapy in the marketplace and has been endorsed by managed care in many cases is the psychoeducational group. Psychoeducational groups are basically classes taught by someone who may or may not be a therapist in which members are directly taught skills and knowledge related to whatever problem or interest the group shares. Psychoeducational groups often impart valuable information related to life skills, but they leave out all the therapeutic elements contained in other theory-based modalities of group therapy: here-and-now interaction and utilization of transferences and group dynamics, exploration of the past for the purposes of insight into the present, affective experience, the exposition of one's dysfunctional beliefs and behavior patterns, and the corrective emotional experience.

## Five Theory-based Group Therapies

Five widely employed group methods (and there are others that are less well known) can legitimately be called theory-based group *therapy:* (1) psychodynamic groups, (2) interactional or interpersonal groups, (3) cognitive-behavioral groups (including Rational-Emotive Therapy), (4) psychodrama, and (5) creative arts therapies. Each of these is based on well-developed psychological theories and requires considerable training.

Psychodynamic therapy groups are based on one or a number of psychoanalytic psychotherapies (Freudian, self psychology, and object relations) which all have in common, as summarized by Rutan (1992) and Alonso (1989), a belief in the existence of psychological determinism and of unconscious processes, as well

---

[1] The preceding comments are not meant to denigrate leaderless self-help groups, which serve a useful purpose and are not pretending to be psychotherapy.

as the assumption that human behavior is dynamic, goal-directed, and epigenetic. The therapy context is seen as a stage for the unconscious reenactment through the intragroup transference(s) of the patient's interpersonal and intrapsychic conflicts and deficits. A stable holding environment is created which enables members to effectively experience and examine their concerns. Psychodynamic therapy gives the patient the opportunity to "use the group forum as a time for comparing and contrasting present character styles and histories of origin in order to help patients learn how their pasts affect the present" (Rutan, 1992, p. 24). Group dynamics are attended to and examined for the information they contain about repressed or projected unconscious material of members. Change is effected by an affective recognition and integration of that which has been repressed, split off, and/or denied.

Interpersonal or interactional group therapy, made famous by Irving Yalom (1995), derives from the psychotherapeutic approach of Harry Stack Sullivan, and focuses on the group as a learning laboratory for members—a context in which they can receive feedback about their maladaptive interpersonal behaviors and experiment with new ones. Emphasis is placed on the here-and-now activation and the corrective emotional experience (Leszcz, 1992), and less on unconscious processes and group dynamics than is the case in psychodynamic group therapy.

Cognitive-behavioral approaches, including Rational Emotive Therapy, developed by Albert Ellis (1992), derive from cognitive-behavioral therapy (Beck, 1976) and employ a number of exercises and homework assignments designed to help patients get rid of or change dysfunctional beliefs that keep them stuck in unproductive behaviors and unpleasant feelings. This approach is future-oriented, cognitive as opposed to affective, and takes little account of unconscious processes of the individual or the group. Responsibility-taking is stressed.

Psychodrama is a powerful modality that, while not as widely used as other forms, is a valuable form of treatment in partial hospitals and inpatient settings. Originated by J. L. Moreno (1946/1977) psychodrama stages within the group the stories and dramas that exist in the mind of members, with members playing a variety of different roles, including those of the "others" in their own lives. The process not only affords insight for members, but provides behavioral role training and fosters spontaneity, which is thought to be a therapeutic force in itself (Nicholas, 1984). Here-and-now interaction and transference reenactments that are not part of the drama and group dynamics are not given much attention in psychodrama. The sociometry of the group, particularly what members have in common, is noted and capitalized upon by the leader ("director") in an effort to harness members' interest and participation in the psychodrama.

Creative arts therapies are group treatments geared toward enhancing the depth and breadth of affective expression on the part of members. Like psychodrama, these methods facilitate nonverbal as well as verbal communication among members. As with psychodynamic methods and psychodrama, the realization and depiction to others of repressed or denied material is thought to be healing in and of itself. The detoxification of shameful and traumatic experiences by representing them creatively is a powerful force for change.

Each of these modalities has found a home in various therapeutic settings. The psychodynamic and interactional models seem to be most prevalent in outpatient environments where the group can maintain a steady and cohesive membership over time, and the uncovering and reworking of defenses can be achieved. Meanwhile, psychodrama and the creative arts therapies are more frequently used on inpatient units where there is a high patient turnover and a consequent need for highly stimulating approaches that will get patients involved with each other and with therapeutic material as quickly as possible without the expectation that core issues will be worked through. Partial hospitals provide an intensive all-day group environment which is suitable for all the modalities, although externally dictated time constraints often impose tremendous pressure on therapists and patients to accomplish more than can be reasonably expected.

## The Need for Research on Theory-based Group Therapy

Students of group therapy need to be equipped with scientific as well as experiential substantiation of the assumption that group therapy works, but as of now such validation does not exist. In their review of the literature, Toseland and Siporin (1986) concluded that empirical studies have as yet failed to demonstrate that group therapy is either more effective than individual therapy in more than 25% of cases or more efficient than individual treatment in more than 31% of cases. There are, however, many problems with the research that Toseland and Siporin reviewed that still have not been corrected: (1) In many studies (as well as in Toseland and Siporin's review) methods of treatment or lengths of time of group treatments are not differentiated (Gazda, 1986). (2) In most studies, comparable populations are not compared vis-à-vis their responsiveness to individual versus group therapy, and the possibility that group therapy, like any intervention, may be differentially effective depending on the population it is trying to help is overlooked. (3) The level and type of training received by the group therapists are usually not spelled out. (4) As Fuhrman and Burlingame (1994) note, investigations into which group therapy fared poorly appear to make no attempt at incorporating or capitalizing on the unique properties of the group format (e.g., therapeutic factors, group as a whole). "They can best be described as individual treatment in the presence of others, with the group format being seen as convenient and economical rather than a powerful interpersonal climate with unique therapeutic factors operative" (p. 7).

Additional problems arise in group therapy research now that managed care has taken such a prominent role in dictating treatment formats. Nowadays group is not only brief but indefinite in duration, with insurance companies often approving only three to five sessions at a time, if indeed they pay for it at all. Attempts to evaluate any treatment's effectiveness are seriously derailed when neither the researcher nor the therapist has control over who receives what treatment and for how long. This is doubly true in the case of group therapy outcome research. The problem is a circular one. Because of the vagueness regarding which group therapies are useful for which populations and why, managed care has little data to work from in deciding when and to what extent to reimburse for group therapy. Conversely, when group therapy practice is controlled by the marketplace rather than by competent clinical judgments, it is increasingly more difficult to obtain accurate information about the efficacy of the intervention.

A continuing problem in therapy research is that clinicians who have access to rich anecdotal and process data about their groups are often reluctant to impose a research agenda on their patients. On the other hand, researchers who are not clinicians often do not have the therapeutic sophistication to target meaningful variables for investigation, nor to create research designs that are neither intrusive nor manipulative of patients.

Despite these obstacles, indeed, because of them, it is important that training programs in group therapy build in solid research components so that we can accurately evaluate the effectiveness of theory-based group therapy as it becomes more widely practiced in the future.

## WHAT IS/WILL BE ESSENTIAL IN TRAINING GROUP THERAPISTS?

In training group therapists for the future we believe the following five components to be essential.

1. *Knowledge of psychopathology,* including not only an understanding of the most recent diagnostic and statistical manuals but also psychodynamic formulations of character pathology and knowledge of *normal individual development* (child, adolescent, and adult); knowledge of and experience doing *individual psychotherapy,* including some familiarity with psychodynamic models (classical, self psychology, and object relations).
2. *Personal experience as a member* of a therapy or training group led by a psychodynamic group therapist.
3. *Didactic and supervised clinical training* in group theory and practice.
4. *Systems approaches,* that is, understanding the group as a system and in the context of the larger organization.
5. *Applications of group therapy* to a wide variety of populations.

We will make some brief comments on each of these components.

## Knowledge of Psychopathology, Individual Development, and Individual Psychotherapy

Ten years ago, it might not have been necessary to specify that clinicians working in mental health settings should have courses in psychopathology, human development, and individual psychotherapy. Today, however, many clinicians working in treatment settings have not been educated at the master's level. Many others have master's degrees in social work from programs that have strongly deemphasized psychiatric social work, or a specialty in marriage and family therapy from programs that bypass individual models of personality, psychopathology, and therapy. In counseling clinicians may emerge from programs that not only do not teach these subjects, but do not require extensive practicum and internship experience. Nurses at the RN or bachelor's level in mental health settings are very apt to have switched into psychiatry from another branch of nursing, and to have had no psychotherapy experience whatsoever.

An important role of the future group therapy educator is that of remedial teacher of psychopathology, individual development, and psychotherapy. As group therapy trainers on-site in hospitals and clinics, we find that many if not most line staff members attending our classes are not able to recognize patients' manifestations of defenses against anxiety, nor are they aware of such phenomena as transference, splitting, and projective identification. We are, therefore, giving them their first exposure to these basic components of interactive process; most are excited by what is taught them and curious to learn more. Needless to say, training in psychopathology and individual treatment should *precede* working as a therapist. Given that it does not, it seems to fall on us as group therapy trainers to integrate into our classes some of the psychodynamic knowledge that students should have acquired in graduate school. There is also some value in learning individual and group modalities simultaneously, before one can become too acclimated to one or the other (Alonso, 1993).

## A Training Group Experience

For many reasons, groups therapy training must begin with an opportunity for students to experience the process as a member. It is impossible to understand how group therapy impacts on patients unless one has been in the "patient" role, whether in an actual therapy group or an experiential training group. While we believe that Tavistock groups are a valuable and important learning experience, we believe trainees should also participate in an activity that more closely resembles a therapy group, in order that they may be exposed to the effects of competent group therapy leadership. While it affords an excellent opportunity to learn about unconscious group processes, participation in a Tavistock group focused on issues of authority does not provide positive models of what a clinician can do to make a group safe, cohesive, and therapeutic for members.

In our training work we have been discouraged to find that clinicians in mental health facilities have neither been members in groups nor had any psychotherapy. A training group or group therapy experience gives clinicians a chance, often for the first time, to see how their own family scripts and dynamics, and their roles in these, are playing out in the here and now of their current life relationships, therapeutic work, and dealings with colleagues. Like all patients in group therapy, therapists in the training or therapy group must get "caught in the act," and learn from the inside out; otherwise they will settle for stunted and intellectualized insights and accept counterfeit material as the real thing.

Therapists will respond to the training or therapy group experience in different ways. Some jump in eagerly, take risks, accept and use feedback constructively, and become group therapy enthusiasts. Others have difficulty relinquishing the therapist role that they have appropriated basically in order to protect themselves against exposing their shame or neediness. While the extremely narcissistically vulnerable therapist must be handled gently by trainers and group therapists, he or she should not be encouraged to pursue group treatment as a line of work, for part of what a group needs to do to become cohesive (and hence therapeutic) is to be allowed to challenge the authority of its leader. One purpose of the training group is to sort out those who are suited to becoming group therapists and those who are more competent working in less threatening treatment contexts.

## Didactic Training and Supervision

Training should include at least 90 hours of didactic instruction according to American Group Psychotherapy Training Program requirements, which can take place either at or away from participants' work sites. This training should include the study of the history of group therapy, a broad theoretical overview for conceptualizing the change process and role of the therapist, ethical considerations, and an overview of research methodology and findings regarding group therapy. Training should be geared to help students work with as wide as possible a range of ego strength, intellectual and verbal ability, age, class level, and ethnic group. Training must include both leading groups alone and cotherapy. The course should also include exposure to group methods other than the primary modality of the student.

While training in any of the theory-based group therapies is valuable, the authors favor the psychodynamic approach. We believe that, while other modalities give group members the stimulation of new ideas, only a psychodynamically based group therapy, appropriately adapted to the educational and developmental level of patients, has the capacity to provide consistent and systematic assistance to the individual. It does this by helping patients begin to understand and resolve the internal conflicts and deficits that are causing anxiety, depression, or other symptomatology, and to make the changes in interpersonal functioning necessary to achieve a healthier adaptation. Psychodynamic therapy has a clear and identifiable model of personality at its base, which assumes that the individual early in life unconsciously cultivates defenses for the purposes of psychic survival. These defenses become part of character structure. Psychotherapy in the psychodynamic group involves the spontaneous and unconscious manifestation by the individual of his or her defenses—both maladaptive and healthy—through interchanges with other members. The function of the group is to allow the individual to see how some of his or her defenses have become outmoded and maladaptive, to understand the function they once served, and to make a choice about whether or not to relinquish them. No other method of therapy gives the patient a mirror of his or her own character and defenses in this way.

Since the psychodynamic approach is highly complex, thorough education in it takes years, a commitment most clinicians and institutions are not willing to make. One of the biggest challenges for group therapy training in the 21st century is to streamline the psychodynamic approach so that it can be both learned and practiced in briefer time spans than it has been in the past.

Direct supervision of one's work is a crucial part of any training. Students should have opportunities to lead with master therapists, to be observed through one-way mirrors or videotape, and to present their work to their peers as well as to a variety of supervisors. According to American Group Psychotherapy Association requirements, trainees must have 300 hours of leadership experience in conjunction with 75 hours of supervision. Not only should supervision assist trainees to better conceptualize how a successful group functions and to develop an increasingly sophisticated set of techniques for implementing their work as therapists, it should also provide a forum for the comprehensive exploration of countertransference issues.

## Learning to Help Any Group to Do Its Work

A major task of the clinician is preparing the group to do its work for maximal therapeutic potential. Part of training therapists involves teaching them how to motivate patients to want to attend the group, as well as coaching patients as to what to expect from and how to use the group.

A major factor in helping members do their work is knowing how much and in what way to warm up a group. Depending on the particular task, population, and setting, groups need various amounts of direct help in getting underway. For example, some groups, such as those in inpatient units, where there is no stability of membership and little natural cohesiveness, require active warming up by means of exercises and other structured beginnings. By contrast, groups that are higher functioning and/or have more time to work together, are better off being allowed to take responsibility for themselves early in the treatment.

We teach trainees in inpatient settings to liberally borrow warm-up techniques from psychodrama and the creative arts therapies in order to get their verbal psychotherapy sessions underway as quickly as possible. One psychodrama technique that is highly effective in warming up a group is action sociometry, in which group members are encouraged to move around the room and to talk face to face to other members. A typical action sociometry instruction might be for members to stand in a particular corner of the room depending on whether they are an oldest, middle, or youngest sibling or an only child, and then to talk to the group members who share the same sibling position. Another might be for each member to put his or her hand on the shoulder of someone he or she might want to get to know better. The pattern of choices is then analyzed, illuminating alliances and connections within the group and identifying "isolates" who will then need to be integrated into the group (Hale, 1981).

The clinician in inpatient and day treatment settings is frequently called upon to lead large groups of 20 or more members, such as multiple family groups or community meetings. The therapist must know interventions that will facilitate the group's accomplishing its task as quickly as possible. In these large meetings the clinician must be able to identify and manage transactions across boundaries, such as between staff and patients, patients and administration, patients and families, staff and families. In the community meeting the therapist helps the group to address a number of activities, such as planning chore assignments, negotiating passes, or resolving conflicts. In the multiple family therapy group the task is to build a sense of community and hope among a disparate group of patients' loved ones who are anxious, confused, discouraged, and ashamed in this exposed treatment setting.

## Systems Approaches

In our work with clinicians in all settings, nothing we teach seems to make a greater impact on them than systems theory does. The systems approach makes perfect sense to therapists who are constantly struggling with a plethora of subsystems on many different levels and wondering why they feel so confused so much of the time. Systems theory liberates them from the need to blame themselves or others, providing a wider frame of reference with which to view problems. Clinicians are pleased to learn that impasses of all kinds—individual, interpersonal, and group-wide—can often be overcome by switching to a larger or smaller level of abstraction and are intrigued by the possibilities in terms of communication that this new way of looking at life presents.

When combined with the concepts of group development (e.g., Bion, 1961) and theme-centered group dynamic theory (e.g., Whitaker & Lieberman, 1964), systems theory is not only an indispensable lens with which to understand group process and group dynamics (Agazarian, 1992), but also essential in aiding students to understand what part the groups they run play in the lives of members, and how the therapy group fits in the context of the system in which it operates (Brabender & Fallon, 1993; Klein, 1992; Klein, Brabender, & Fallon, 1994). A systems approach also provides an avenue to understanding complex communications in the group (Agazarian, 1992; Nicholas, 1984).

## Applications of Group Therapy Theory and Practice

A central thesis of this chapter is that generic group therapy training that is psychodynamically based will equip psychotherapists to run excellent therapy groups in any setting. In addition to the standard generic training, however, students must receive specific instruction on how to do group therapy with people with addictions, eating disorders, chronic physical and mental illness, post-traumatic stress disorder, homelessness, crisis situations and other special problems. One of the group therapist's basic functions is to assess the composition of the group at hand and design his or her approach to accommodate the needs of that particular constellation of patients. If the clinician is adequately trained in the generic model we are suggesting, he or she will be able to discern the balance of nurturing versus challenge that would be optimum, whether the group consists of elderly organic patients, teenage drug addicts, or a combination of both (it happens!). The therapist will be able to determine when to be directive or nondirective, when to show himself or herself openly in the group and when to remain "opaque," when to use warm-ups or when to let the group struggle with its own beginnings, and when to let the anxiety build or reduce it. A properly trained group therapist will automatically gear the level of his or her vocabulary and conceptualization to meet that of the patients and will know when to use nonverbal techniques to supplement or promote verbal interaction.

## PLANNING FOR THE FUTURE WITHOUT LOSING THE PAST

Group therapy education for the 21st century must be both remedial and progressive. In addition to getting into institutions to train therapists who are already doing group therapy half-heartedly and incorrectly, we must also train the group therapists of the future. Training these group therapists will involve equipping clinicians to design and run groups that provide maximal therapeutic opportunities for a wide variety of patients within very short time periods.

Training group therapists to deal with the economic and time constraints of mental health delivery systems in the future does not mean that they should not be taught to do long-term group therapy as well. Contrary to the predictions of doomsayers, long-term psychotherapy will survive. The chronically mentally ill will continue to need it, and vast numbers of others will feel strongly enough about its benefits to seek it, even though they must pay out of pocket. The format for such treatment, however, will increasingly be group, and the setting private practice. Group therapy is an economical treatment that compensates the practitioner reasonably well, at low fees for the members. As a treatment it is as powerful, if not more so, than individual treatment and may emerge as the fuel that keeps private practice and long-term therapy alive in the 21st century.

# REFERENCES

Agazarian, Y. (1992). Contemporary theories of group psychotherapy: A systems approach. *International Journal of Group Psychotherapy, 42*(2), 177–203.

Alonso, A. (1989). The psychodynamic approach. In A. Lazare (Ed.), *Psychiatry: Diagnosis and treatment* (2nd ed., pp. 37–58). Baltimore: Williams & Wilkins.

Alonso, A. (1993). Training for group psychotherapy. In A. Alonso & H. Swiller (Eds.), *Group therapy in clinical practice* (pp. 521–529). Washington, DC: American Psychiatric Press.

Beck, A. (1976). *Cognitive therapy and the emotional disorders.* New York: International Universities Press.

Bion, W. (1961). *Experiences in groups.* London: Tavistock.

Brabender, V., & Fallon, A. (1993). *Models of inpatient group therapy.* Washington, DC: American Psychological Association.

Budman, S., Soldz, S., Demby, A., Feldstein, M., Springer, T., & Davis, M. (1989). Cohesiveness, alliance and outcome in group therapy: An empirical examination. *American Journal of Psychiatry, 52,* 339–350.

Dies, R. (1992). Models of group psychotherapy: Sifting through confusion. *International Journal of Group Psychotherapy, 42*(1), 1–18.

Ellis, A. (1992). Rational-emotive and cognitive-behavioral therapy. *International Journal of Group Psychotherapy, 42*(1), 63–79.

Fuhrman, A., & Burlingame, G. (1994). Group psychotherapy: Research and practice. In A. Fuhrman & G. Burlingame (Eds.), *Handbook of group psychotherapy: An empirical and clinical synthesis* (pp. 3–40). New York: Wiley.

Gazda, G. (1986). Discussion: When to recommend group treatment: A review of the clinical and research literature. *International Journal of Group Psychotherapy, 36*(2), 203–206.
Hale, A. (1981). *Conducting sociometric exploration: A manual for psychodramatists and sociometrists.* Roanoke, VA: Ann Hale.
Klein, E. B. (1992). Contributions from social systems theory. In R. Klein, H. Bernard, & D. Singer (Eds.), *Handbook of contemporary group psychotherapy* (pp. 87–124). Madison, CT: International Universities Press.
Klein, R. H. (1979). A model for distinguishing supportive from insight-oriented psychotherapy groups. In W. G. Lawrence (Ed.), *Exploring individual and organizational boundaries* (pp. 135–152). London: Wiley.
Klein, R. H. (1993). Short-term group psychotherapy. In H. Kaplan & B. Sadock (Eds.), *Comprehensive group psychotherapy* (3rd ed., pp. 256–269). Baltimore: Williams & Wilkins.
Klein, R. H., Bernard, H., & Singer, D. (1992). Introduction. In R. Klein, H. Bernard, & D. Singer (Eds.), *Handbook of contemporary group psychotherapy* (pp. 1–23). Madison, CT: International Universities Press.
Klein, R. H., Brabender, V., & Fallon, A. (1994). Inpatient group therapy. In A. Fuhrman & G. Burlingame (Eds.), *Handbook of group psychotherapy: An empirical and clinical synthesis* (pp. 370–415). New York: Wiley.
Leszcz, M. (1992). The interpersonal approach to group psychotherapy. *International Journal of Group Psychotherapy, 42*(1), 37–62.
Moreno, J. (1946/1977). *Psychodrama,* vol. 1, 4th ed. Beacon, NY: Beacon House.
Nicholas, M. (1984). *Change in the context of group therapy.* New York: Brunner/Mazel.
Ormont, L. (1993). Resolving resistances to immediacy in the group setting. *International Journal of Group Psychotherapy, 43*(4), 399–418.
Piper, W., & McCallum (1994). Selection of patients for group interventions. In H. Bernard & R. MacKenzie (Eds.), *Basics of group psychotherapy.* New York: Guilford Press.
Rutan, S. (1992). Psychodynamic group psychotherapy. *International Journal of Group Psychotherapy, 42*(1), 19–35.
Toseland, R., & Siporin, M. (1986). When to recommend group treatment: A review of the clinical and research literature. *International Journal of Group Psychotherapy, 36*(2), 171–201.

Whitaker, D., & Lieberman, M. (1964). *Psychotherapy through the group process.* New York: Atherton.

Yalom, I. (1995). *The theory and practice of group psychotherapy* (4th ed.). New York: Basic Books.

# 13

# Changes in the Professional Marketplace:

## A Personal Odyssey

*William Hausman, M.D.*

The past few decades have been marked by major changes in the patterns of organizational structure and leadership in professional as well as business and financial institutions. These changes, in turn, have affected the activities of both administrators and professional workers. As with all change, the effects on those involved, whether in the role of staff or recipient of services, has varied greatly. It is axiomatic that "all change is loss." But it is also true that the crisis of change can lead to significant opportunities (Hirschowitz, 1990). These opportunities may include more efficient organizations and more functional professional roles. They may or may not benefit the system's clients or the nation's well-being.

This chapter will examine some of the recent changes in a particular field, psychiatry, from the perspective of an individual whose professional career has spanned more than 4 decades and has included work in three very different types of settings. That career has included marked changes of venue on two occasions, the most recent occurring about 9 years ago. The chapter will also examine the changing patterns of psychiatric hospital governance in recent years and some of their effects on the delivery of care. These changes are examined in the context of the theories of organizational systems propounded by Miller and Rice (1967). Their work focused on the nature of leadership, organizational boundaries and their transactions, the concept of "primary task" (defined by Miller and Rice, 1967, as the task that an organization must perform if it is to survive), and authority in organizations. In this examination of changes in facilities for treating the psychiatrically ill, the economically driven forces affecting leadership and their implications for the primary task in these settings will be explored.

My vantage point is from work as a *locum tenens*\* psychiatrist for brief periods of time in a variety of different psychiatric settings during the past 9 years. The role of the locum tenens, a unique and recently burgeoning form of practice pattern, will be described. This experience has permitted me to examine some of the more recent changes in the institutional practice of psychiatry and their consequences for staff and clients. Although the use of "locums" to fill vacancies in various types of institutions has been part of the practice scene for many years, the rapid emergence of new forms of treatment facilities, coupled with early physician retirements, has generated novel practice options. This change has helped to create new agencies that contract with professionals to fill the vacancies created by rapid expansion of facilities and turnover of clinicians.

## THE OBSERVER AS HISTORIAN

My psychiatric background is germane to these observations on recent change. I entered psychiatry following graduation from

---
\*Term applies to professionals (e.g., physicians) hired on a temporary basis to fill an open position.

medical school in 1947. My initial training was in two venerable psychiatric institutions, one a public hospital, Worcester (Massachusetts) State Hospital, and the other a private clinic, The Institute of the Pennsylvania Hospital (Philadelphia). Both systems had served as treatment and training centers for over a century. They afforded early exposure to then contemporary practice in institutions steeped in the history of American psychiatry. My subsequent clinical experience included duty as a commissioned army medical officer, including a tour as division psychiatrist in combat. The battlefield experience and subsequent army assignments introduced me to the notion of social psychiatry, that is, "the interrelationship between the sociocultural environment and the individual" (Kiev, 1969). This concept and the related but oft-disputed notion of community psychiatry (Dunham, 1969) were important to me for perspective and for career directions.

After retirement from a military career, I spent the next 21 years in two academic institutions, 11 of them as a medical school department head and chief of psychiatry in a university teaching hospital. These positions involved work as an educator, as a consultant to a series of community organizations in an evolving "new town" (Columbia, Maryland), as an academic and clinical services administrator, and as a clinical psychiatrist.

My work in the group process area during the early development of the organization that has become the A. K. Rice Institute (AKRI) started shortly before retirement from the Army, and that association has continued until the present. That career pattern began with a Tavistock Institute group relations seminar in England and has included many conference and organizational roles, mostly in the early period of the AKRI's development in the United States. Close association with A. Kenneth Rice, who led the new enterprise with a group of British and American colleagues, made the experience with the early development of the "Tavistock" work in the United States particularly valuable. More recent association with the Levinson Institute, an organization that conducts seminars for business leaders and offers psychological consultation to businesses, has complemented my work with the AKRI. While the AKRI has emphasized the behavior of the organization and its component groups, the Levinson Institute, under the leadership of Harry Levinson, has focused more on individual roles in the organization. The approach of both institutes has stemmed from psychoanalytic theory.

The area of psychiatry that I found most attractive and most consonant with my personal philosophy throughout my career has continued to be social psychiatry. This approach to understanding the basis for behavior and treatment was reinforced, as noted earlier, by my early experience in the combat arena of the Korean War. It was later bolstered by my work as a research administrator and clinician in other Army settings and by my contact with research scientists around the world while serving as Director of Behavioral Science Research in the Army Surgeon General's office. It was further supported by my experiences with the AKRI and the Levinson Institute.

My retirement from full-time work marked a hiatus in a professional career that had by then spanned about 40 years. During those years the field of psychiatry had changed markedly. In the 1940s and 1950s psychoanalysis was a preferred treatment not only for neurotic patients, but also for some psychotic individuals whose families could afford the high costs and experimental nature of this approach. Electric shock, insulin coma, and lobotomy were available in both state and private hospitals as the principal approaches to the treatment of psychoses. Sedation of psychotic excitement was accomplished with warm baths, cold sheet packs, seclusion, and traditional sedative medication. By the mid-1960s, the development and use of newer types of antipsychotic, antidepressant, and anxiolytic medications had revolutionized the treatment of many types of psychiatric disorders. These agents, and the introduction of the concepts of social and community psychiatry and the related notion of deinstitutionalization at about the same time, produced remarkable reductions in the census of state and county psychiatric hospitals. By the 1970s psychiatric patients were seen in the acute phases of their illness largely in city and teaching hospitals and in some private institutions, often supported by private insurance programs and by the federal programs Medicare, Medicaid, and Champus.

By the 1980s several corporate groups of for-profit psychiatric hospitals and clinics had emerged and were actively competing for what was seen as a lucrative market in (mostly acute and subacute) psychiatric care of adult, adolescent, and child patients. With the introduction of several new families of psychoactive drugs, the focus of psychiatric treatment rapidly shifted to medical therapies. Several academic centers aggressively challenged the usefulness of dynamic therapies and the field rapidly became dominated by medical approaches. Research in psychiatry became heavily directed toward biological aspects of mental disorders. The for-profit and most other clinical centers largely emphasized treatment with medications. The increasing costs of psychiatric treatment and, to some extent, the ambiguity of descriptive language about illness and patient response caused insurers to implement rigorous management systems, starting in the mid-1980s. These paralleled similar approaches to treatment of other types of medical disorders. The patient management systems have encouraged the aggressive use of medications in psychiatric treatment largely because they shorten expensive hospitalizations and often reduce the frequency of hospital admissions.

In effect, history has caught up with the field of psychiatry, reinforced by powerful economic and political forces. The earlier approach to treatment incorporated a variety of organic approaches to distressing chronic disorders. The advent of psychoanalysis introduced new hope for some patients and interested many clinicians. New approaches to therapy broadened the range of available treatments and increased the number of clinicians in several disciplines who offered their services to less disturbed psychiatric patients as well as to selected groups of psychotic individuals. These treatment approaches were expensive because of the number of sessions required. The nature of the disorders were varied, and diagnostic criteria were less than precise. In the last few decades, efforts have been made to better classify and describe psychiatric illnesses so that more definitive research can be conducted.

The rapid movement toward a medical approach to treatment has coincided, to a large degree, with the development of newer types of hospitals. In turn, this has led to intense competition for those patients that are privately insured. Federal programs, like Medicare and Medicaid, also entered into the psychiatric system, each with different requirements and limitations for support of treatment programs. In their concern for justifying support of their programs, clinicians and hospitals made use of the newly updated systems of nomenclature to justify billing for patient care to the various third-party payers. This in effect created the potential of corrupting the "scientific" new diagnostic criteria by using them to justify support of inpatient or outpatient treatment of psychiatric patients. Rules set up by various insurers and federal agencies to define treatment criteria are often used as yardsticks for billing practices. To reduce rapidly escalating costs, these third-party payers also set time limits for treatment, sometimes manifesting minimal concern for the patient. Thus, traditional collaboration in seeking optimal treatment often has been replaced by an adversary system.

The climate of the field changed markedly as new patterns of practice rapidly evolved. For clinicians who had found psychoanalytic and related dynamic approaches to treatment interpersonally and intellectually challenging, the new directions and their economic concomitants have often been discouraging. While medical therapies have offered real hope to some depressed, phobic, and obsessive individuals, the valuable interpersonal skills that had often made medical treatments unnecessary were devalued (McIntyre, 1994). In many cases severely psychotic individuals were helped by these new medical approaches, but economic pressures often resulted in early discharge of patients who might have profited from longer hospital stays. Access to outpatient and outreach systems to support these very disturbed individuals after discharge has varied greatly among communities. As a result, the same type of psychotic person might be followed intensively in one city and be virtually without support in another. Recent cutbacks in welfare funds seems to have disproportionately affected the care of these vulnerable patients in most states.

My retirement at the relatively young age of 62 fell into the period of adult development defined by Levinson et al. (Levinson, Darrow, Klein, Levinson, & McKee, 1978) as that of late adult transition, a period where work is shaped by legacy and the opportunity for contributions to humanity. The decision for retirement was influenced by the changes in practice patterns and in the priorities of the professional arena as just described. Consistent with my stage of adult development I sought a greater sense of freedom to pursue a more gratifying professional career, free of the constraints of the university and other formal practice settings. Having left full-time employment, the clinician must decide whether to engage in activities outside his or her career field or to find alternative ways of maintaining professional involvement. The remainder of this chapter will deal with my choice and the experiences and changing views of psychiatry that have followed my decision.

## THE WORLD OF THE LOCUM TENENS

The clinician who is in the midst of a career situation rarely gives much thought to alternative options. However, as one approaches the point of leaving an established role, other choices begin to present themselves. After correspondence with an agency offering locum tenens opportunities, I completed the necessary documentation and was accepted by that agency. I found that the types of assignments available varied widely, and that the clinician is free to accept or reject a given assignment. In the past 9 years my locum sites have included a Veterans Administration hospital, several state hospitals, community not-for-profit and for-profit hospitals, a regional outpatient program, and a large HMO organization. My roles have varied from being ward physician in chronic psychiatric treatment programs, to staffing acute treatment wards and adolescent units, to working in outpatient care and consultation-liaison activities.

My patients have included retarded and psychotic adults, delinquent adolescents, depressed workers, and long-term nursing home residents. Their diagnoses have covered the handbook of psychiatric nomenclature. Treatment options have been broad and at my discretion. Where the type of disorder is relatively new to the locum's clinical experience, a colleague in the facility always has been available to guide the treatment approach and to suggest appropriate literature on the subject. In all institutions consultation has been available for the assessment and treatment of medical and other nonpsychiatric aspects of a given case. Thus, what had seemed at first to be a formidable task of having to rapidly walk from one treatment abyss to the next has turned out to be manageable and interesting as well as being an opportunity for continued learning. As important to me at this stage of my life, I have had the opportunity to draw upon my long and diversified career in my work with therapists and administrators of the different facilities in which I have worked. Most of these individuals are much younger than I, and their clinical experience dates from recent years. Their interest in gaining a historical perspective of psychiatry from a visiting clinician is keen. In turn, they have helped me to better hone my skills in the modern hospital setting. Although aware of the importance of keeping current on knowledge of the newer psychopharmacological treatments, my major emphasis has been on a more eclectic approach, integrating medical and psychological treatment of the patient. With the administrators of the hospitals, my experience as an organizational consultant has been valued and has given me increased credibility. Similarly, my experience in academic centers has been of interest to those clinicians whose careers have had a more academic focus. My concern for the patient and for teaching younger staff, true to my stage of adult development, has enhanced my personal experience as a locum and seems to be rewarding to those with whom I have worked.

## CHANGES IN THE PRACTICE ARENA AS SEEN BY THE LOCUM

Just as the patterns of psychiatrists' practice have changed in the past decade, so has the nature of psychiatric facilities. These changes are a product of the previously described patterns of diagnosis and treatment. They are also strongly determined by the economics of health care and by the emergence of organizations designed to gain maximum profits from the business of caring for the mentally ill. Indeed, the largest growth of new facilities is in the for-profit sector. This type of institution is often distinguished by a sharp demarcation between the administrative staff, led by a highly paid, often nonclinician, chief executive officer, and therapists who care for the patients. The "front office" is large, well staffed, and equipped with state of the art computers. Marketing is an essential function, as is the financial office, where the insurance status of patients is closely tracked. Front-office staff maintain close communications with their counterparts in the corporate offices, usually in another city. In some of these facilities critical decisions about the hospital seem to be closely controlled by the corportion. In one such hospital there was a dramatic contrast between the high-tech front office suite and an antiquated system of typing and filing medical records.

In some for-profit hospitals a few individuals have been assigned a role that bridges the front office and the clinical services. They are designated "clinical care coordinators" or "therapists." In these roles they seem to be expected to represent the concerns of management more than those of the patients or the clinical staff. They "coordinate" length of stay and planned dates of discharge, with a close eye on the insurance status of the patient and the bed occupancy rate of the hospital. These staff members seem to have been placed in a position where they must regularly feel the pressure of being in conflict of interest. While it is true that the physician must order a discharge, these staff are expected to have the responsible clinical staff members understand the economic consequences for the hospital of decisions about discharge or retention of patients. (On one occasion a person in this role who attempted to encourage my extending the period of hospitalization of a well-insured recovering patient suggested that she might be fired if the patient was not retained.) Similarly, although the physician in most facilities must concur before an admission is accepted, it is the marketing office that orchestrates the process and provides much of the information on which the decision is made. These and similar practices are changing in recent years as the legal system has caused extensive modification of the way in which these for-profit facilities are managed, as I will discuss.

In the past few years the medical insurance industry has stepped up its role in "managed care," effectively reducing hospital stays at all types of medical facilities that depend on insurance payments for patient care. Managed care has placed additional constraints on clinical decision making in these hospitals and on the administrator to maintain the profitability of these facilities.

The locum, who is working at the facility for a relatively brief period, soon recognizes that it takes time to "learn the ropes." He or she works alongside of the regular physicians, many of whom are beholden to the administrators for their income and sometimes their career. While the physicians working in these for-profit facilities usually have accepted these conditions as "going with the territory," their morale suffers and their investment in the work is often low. One physician, the medical director, is appointed and largely paid by the hospital and also maintains a practice there. While assuming leadership of the medical staff, he or she may be viewed with distrust by the other physicians who see that doctor as having a greater investment in the priorities of the front office than in those of the professional staff or patients. Some of these facilities have a preponderance of Medicaid or Medicare patients. The facility cannot function without physician staff, and would not be reimbursed by third-party payers without having doctors assigned to each patient in order to be reimbursed by third-party payers. Doctors' Medicaid and Medicare fees are limited at this time, so the hospital must pay the all-essential doctor out of hospital operating income if it is to retain his or her services. This applies also to the regular medical staff members, many of whom have relied on such supplemental income to support their continued work at the hospital.

The result of these arrangements is that the for-profit system must pay the cost of highly salaried hospital directors, medical staff supplements, locum fees, the income of corporate staff, and dividends to shareholders, all with funds received for patient care. These facilities rarely sponsor outpatient or follow-up care as that approach seems to be less lucrative for the for-profit enterprises. Instead, some of the more enterprising organizations have had a network of outreach offices, manned by marketing-oriented "therapists." It should be noted that in the past few years such practices have been the basis for legal action by the attorneys general of several states and have prompted some large and highly publicized lawsuits. These have caused some of the largest companies to make significant changes in the management of their psychiatric hospitals, to close some and sell off others (R.B.K., 1994). It is hoped that these changes will help create a more therapeutic climate in the for-profit facilities.

It is not surprising in this situation to hear complaints of overwork on the part of ward nurses, physicians, and other clinical staff members. Many of these facilities are located in small communities where labor costs are relatively low and the staff is likely to have few alternatives to working at that center. In my experience the staffs are devoted to patient care and usually demonstrate a high level of professionalism. They easily form an affinity with the locum, with a common concern for the care of the patient.

In the nonprofit community hospitals the economic dynamics are somewhat different. Patient input is highly dependent on the regular medical staff's practice patterns and their relationship with colleagues in the region. The medical director is usually elected by the medical staff. In recent years some of these hospitals have expanded their marketing function, usually through media advertising, as they compete with for-profit and other nonprofit facilities. While usually not part of large corporate entities, they sometimes contract with national or regional organizations for management services. Because most of these facilities have psychiatric wards as part of a much larger medical-surgical facility, many carry substantial debt loads from the acquisition of high-technology medical equipment and new or upgraded hospital wards. The high cost of this debt and the need to maintain a positive balance sheet has increasingly resulted in competition for leaders with good business credentials. This in turn results in higher leadership salaries and large increases in front-office expenditures. The tendency to bloat this function of the hospital creates a situation similar to that already described for the for-profit ventures. These settings more often provide facilities for outpatient diagnosis and care or develop close working relationships with one or more practice groups in the community. Like other types of medical facilities, practice in these settings has also been constrained by recent increases in the "managed care" operations of medical insurers.

While my sample of health maintenance organization (HMO) settings is more limited, these facilities tend to operate on a different basis. In some, more senior medical staff are shareholders in the enterprise and so are motivated toward keeping the cost of care as low as possible. At the same time, a marketing effort must be carried out with business and governmental organizations in the area to secure contracts for medical services to the employees and staff of those groups. In order to succeed in the marketplace not only is the low cost of care important, but also a reputation for high quality treatment and good service. In most HMOs fee-for-service payments, typical of other types of private health care organizations, are replaced by capitation systems. This offers an incentive to the physician-shareholders to place greater emphasis on preventive care and less on costly procedures and laboratory work. The patients and the organization both gain from reduced administrative procedures and costs around the registration and billing processes. For the locum, concern for the patient is shared by professional colleagues of all fields, with less of the tension between professional and administrative staffs of the facility than was described with the for-profit hospitals. While my brief experience in this type of facility was positive, some HMOs have been criticized for excessive concern about cost-saving to maintain bonuses for the staff. These decisions are sometimes at the expense of sufficient hospital time to assure full patient recovery. In this sense their approach to economic incentives works opposite to that of the for-profit hospitals. As prototypical "managed care" organizations, some HMOs' practices have recently led to legislative control of these practices by national and state legislatures in an attempt to protect consumers.

Although the HMO was originally developed as a nonprofit type of fiscal structure, a number of organizations have made use of the HMO concept in the framework of a for-profit organization. In one case of such a mid-size HMO (*San Francisco Chronicle*,

May 5, 1994) the CEO's recently reported income was higher than that of the CEOs of a number of national Fortune 500 corporations.

The state hospital is a very different sort of setting. Here the financial constraints are the lack of public funds allocated to the care of a scapegoated part of the population. When states' budgets are strained, as has been true in the past few years, it is politically safe to reduce funds for the mentally ill. Organizations like the National Alliance for the Mentally Ill (NAMI) have emerged in the past decade to integrate the previously ineffectual efforts on behalf of the mentally ill. Despite NAMI's attempts to spotlight the plight of this needy group, the public has generally lent a deaf ear to its pleas. Ironically, the "progressive" deinstitutionalization policies of the 1960s helped to create the current dilemma as state facilities closed units, reduced staff, and put the savings to other use. In fact, many of those patients "freed" from public mental hospitals have been forced to live on the streets with little help available to them for professional care or basic sustenance. This phenomenon is more pronounced in this country than in England, which has a very different system for supporting the treatment of the mentally ill (Cohen, 1994). The state hospitals have become "revolving doors" for these mentally handicapped citizens. Budget reductions over the years have led to crowded wards, understaffing, and reduced community support programs. In many states tight budgets preclude effective staff professional development. Recruiting of psychiatrists and other professional staff members is hampered by state civil service regulations and by limited salaries, placing these deserving facilities in a poor competitive position with private institutions in the same areas. In several state hospitals the recruitment of a medical director was delayed (sometimes for years) by these limitations. The political base of these institutions has tended to foster interdisciplinary struggles, which, in some settings, has also affected recruitment. In such a climate the locum is responsible for a large group of needy patients under the daily care of undermanned and often undertrained nursing staff. Despite these severe constraints, when comparing the facilities in which I have recently worked with those experienced 4 decades earlier, the present state hospitals show an improved and less depressing, albeit far from optimal, therapeutic environment. In a number of settings new facilities had been built in recent years which were more modern and better designed for today's treatment programs. The range of treatments available are greater and there is better linkage with community mental health programs that assure stronger continuity of care for the discharged patient than had been true in the past.

I have had some brief experience as a locum in a Veteran's Administration (VA) hospital since retirement. These settings share some of the problems of the state hospitals in the complexity of their bureaucratic structure. However, they are generally better funded, thanks to the political power of veterans organizations. On the other hand, the patients in these facilities are highly motivated by the pension benefits that accrue to the medically disabled veteran, and much of the treatment climate of these hospitals is affected by pension-seeking and pension-retaining strategies (Satel, 1994). In my limited experience in the VA system it would appear that the politics of interdisciplinary competition are alive and well in these facilities, often at the expense of optimal staffing of the clinical services.

I have not worked as a locum tenens in free-standing private nonprofit psychiatric institutions or in teaching hospitals. Prior to retirement and after leaving the Army, my experience was limited to academic institutions in two settings, and I did some consulting in other academic departments of psychiatry. These settings have been in the process of change during the 2 decades that I spent in them, and this change continues. Some teaching hospitals have been taken over by for-profit medical companies, which has undoubtedly changed the working climate of these institutions and may well be expected to temper the idealism of medical students and professional trainees.

Reflecting back on this potpourri of professional experiences since "retirement," I find that the work roles that have been most professionally gratifying were in several community hospitals and the HMO, followed closely by some state hospitals. As the preceding descriptive comments must convey, the least gratifying professional environment for me (and likely for full-time medical colleagues) has been in the for-profit corporate setting. This is not surprising, considering my personal history and my stage of adult development. My background, with its emphasis on community psychiatry, psychological aspects of treatment, and patient-centering of the treatment process, all bias me strongly toward those programs and institutions that emphasize these elements of care. My work in the military and in nonprofit teaching hospitals and my close alliance with students in university settings have reinforced my humanistic and liberal attitudes about the work that doctors and medical institutions do. My experience of taking on this new career brings home to me the psychological impact of this stage of adult development on the work that I do in my "retirement" years. The competition for status and economic advantages of earlier years has been supplanted by a need to put my professional legacy to work as a mentor and to actively contribute to the needs of my patients. The economics of retirement also permits me more freedom in where and when I work. Happily, the locum tenens role offers a great potential to satisfy these changing personal needs at a time when some of the world of professional practice seems to be moving in an opposite direction.

## LEADERSHIP IMPLICATIONS

In examining the experiences described here, I have been aware of significant changes in the leadership of psychiatric treatment services during the past few decades. The clinician-as-leader, in many situations, has been supplanted by the business-of-medicine leader. This has been most evident in the for-profit facilities, where the primary task has changed from the care of the patient in need to that of maintaining profitability for the corporate owners of the organization. In public hospitals the clinician-leader has been largely replaced by the bureaucratic expert who seeks to have the facility survive in the face of budget cuts and scapegoating legislators. The not-for-profit community services also are concerned at satisfying increasingly complex rules for operating their services and thus fall between the state hospital's operating constraints and those of the medical corporations. Leadership in these settings must recognize the conflicts between the fiscal and bureaucratic constraints on the system, including the advent of "managed care" insurance programs, and the morale of staff and well-being of patients. In many of these community settings the larger hospital staff, administrative and professional, has little knowledge of or interest in the work of the mental health organization, so that necessary staffing augmentations, maintenance, and variances in hospital rules are obtained with difficulty by the psychiatric unit leadership.

The locum tenens is in a unique role as regards leadership. Although his or her role as physician (and sometimes the only psychiatric physician) is essential to the clinical and fiscal operation of the unit, the locum is only a temporary employee. Thus, the management of the locum's role is a significant challenge to the unit's leadership. The other staff of the unit must adapt to a variety of clinical approaches, to a range of personalities, and to different leadership philosophies by the various assigned locum tenens psychiatrists. The staff are obviously affected by these differences, engendering responses from indifference to dependency to hostility by clinicians of mental health-related disciplines in that organization. The leader of the unit must be prepared to add this problem to the myriad of others required for leadership of today's mental health organization.

Although a temporary person in the organization, the locum has the opportunity to exercise leadership in most mental health facilities. This is a result of the authority that has been traditionally inherent in his or her role in the medical organization. It is also related to the fact that very few decisions about a patient's care can be made without the psychiatrist's concurrence. Admissions, discharge, consultations, medications and their changes, and the use of procedures like restraint all require a doctor's order. Beyond these statutory aspects of the psychiatrist's role, the experience of a seasoned clinician, as is true of many locums, permits that person to exercise functional leadership in many clinical settings. The challenge for both the locum-as-leader and the permanent leadership of the facility is to quickly put in place a relationship that exploits the potential of that situation. In my own experience, when that interaction has been effective the work has been very satisfying to me and appears to have had value for most or all of the unit's professional staff. In such a setting the ultimate winner is the patient.

The locum tenens, embodying a relatively new role in the medical organization, is but one of several medical professionals who are assigned roles in the hospital or clinic on a temporary basis. As such, these professionals may represent a phenomenon somewhat akin to the temporary task team of the modern commercial aircraft who are assembled to carry out a highly specialized task. When the task has been completed (e.g., the aircraft arrives at its destination), the team is disbanded and another such team may be formed for the next leg of the aircraft's schedule. In the modern hospital organization, some of the physicians may be locums, some nurses come from a nursing pool (or may also be locums), and other staff may be brought in during heavy admission periods. Managing such an organization represents a unique and potentially productive challenge for medical service leadership, somewhat parallel to the task-centered leadership of some modern technological businesses where the temporary task team may represent a creative management tool that exploits the skills of its members for a defined project in a limited time period.

## SUMMARY AND CONCLUSIONS

While patterns of professional practice in psychiatry have changed impressively over the past decades, the pace of change has never been as striking as in the past 15 years. Much of this accelerated change is a result of significant research and therapeutic developments that have dramatically altered the treatment of the mentally ill. Interprofessional politics and economics have also contributed to change as practitioners of several mental health disciplines have vied for income and status. However, the largest factor in this revolution has to do with the economics of the health care industry. During this period several large corporations have emerged to exploit the burgeoning market for mental health services and the public and private funding systems developed since the 1960s to support the care of the mentally ill. This movement followed the reduction of funds for public care of the psychiatrically ill. The consequences of these events have significantly impacted the patients, the cost of health care, and the morale of professionals working in the field.

In the spirit of the 1980s, greed and the uncontrolled striving for profitability has been no less evident in the mental health treatment arena as elsewhere in the American economic scene. One consequence of this trend has been the proliferation of corporate for-profit treatment facilities. This has affected not only competition for patient care income but also organizational structures and professional careers. The professional's commitment to the care of patients has been undercut by his or her subordination to corporate priorities, with clear, and sometimes profound, effects on the work climate.

A. Kenneth Rice (personal communication, June, 1968) has suggested that the organization's leader embodies the task. This concept has required reexamination in the new world of corporate professional work. If that thesis is true, then it would appear that the primary task of the for-profit hospital has changed from that of patient care to fostering the interests of corporate officials and their investors. Doctors, nurses, and others experience dissonance in their attempts to reconcile professionally engrained philosophies with the demands of work as professionals in many of these new treatment settings. The patient's needs are also at risk of being subordinated to other considerations.

Of the other facilities described from the locum's vantage point, the state hospital also has had difficulty fulfilling its social mandate, but for quite different reasons. The political process in most states has often left these important treatment facilities without adequate financial support for carrying out their missions. Hospital and mental health system leadership in these settings may be inordinately sensitive to political masters. Although state hospital and for-profit facilities are diametrically opposite on the socioeconomic ladder, both are marked by upward-directedness of leaders who are beholden to higher authorities. In turn, the commitment of those authorities is often less to the needs of the client than to those of powerful economic, administrative, and political systems.

The other types of facilities experienced by me were an HMO and some nonprofit community hospitals. Both of these types of programs represented well-staffed and strongly patient-oriented treatment settings. The relationship between administration and professional workers, while not free from problems, was relatively functional in these facilities, and in general there was good collaboration in the task of caring for patients. In the community hospital, in particular, the linkage to community resources was generally good, contributing effectively to continuity of care. In the well-functioning HMO continuity of care is provided within the HMO inpatient-outpatient system.

In examining the profound changes in psychiatric institutions in recent years, the notion of crisis comes into focus. If crisis is, as earlier suggested, an amalgam of opportunity and loss, then these changes must be examined in that context. Opportunity comes both for those patients whose prospects for a better life have been enhanced by new treatment methods and for the medical entrepreneurs and investors who have found the business of psychiatric care to be profitable. Loss is evidence by demoralized staff in several types of facilities, in the reduced resources available to public hospitals for the care of indigent and uninsured mentally ill persons, and in increasing costs to society of treatment programs designed for optimal profits and high administrative overheads. Above all, the loss is also borne by some patients whose needs for help are exploited by some parts of the health care industry.

In a world characterized by rapidly shifting national and professional priorities and by significant changes in roles and institutions, the locum tenens role represents a special opportunity for the professional to become a thoughtful observer, and sometimes a leader in a variety of treatment settings. He or she not only brings to bear his or her legacy in a complex and changing field, but also has the chance to view some of the changes in the field from the vantage point of a seasoned observer. In doing so, the locum notes the trends in the field that represent significant advances in the ways of caring for the mentally ill. He or she also has the opportunity, from a neutral vantage point, to view the foibles of practice that have presented themselves in the field of psychiatry in recent years. The unorthodox concept suggested by the work of R. D. Laing (Laing, 1970; Laing & Esterson, 1970) of being sane in insane places may well take on new meaning in this context.

## REFERENCES

Cohen, C. (1994, February 4). *Homelessness in New York and London. Psychiatric News,* p. 18.

Dunham, H. W. (1969). Community psychiatry: The newest therapeutic bandwagon. In A. Kiev (Ed.), *Social psychiatry* (Vol. 1, chapter 9). New York: Science House.

Hirschowitz, R. G. (1990). *Managing stress and change.* Belmont, MA: The Levinson Institute.

Kiev, A. (1969). Introduction. In A. Kiev (Ed.), *Social psychiatry* (Vol. 1, p. 33). New York: Science House.

Laing, R. D. (1970). *The politics of experience.* New York: Ballantine Books. Chapter 3.

Laing, R. D., & Esterson, A. (1970). *Sanity, madness and the family.* Harmondsworth, England: Penguin Books.

Levinson, D. J., Darrow, C. N., Klein, E. B., Levinson, M. H., & McKee, B. (1978). *The seasons of a man's life.* New York: Knopf.

McIntyre, J. (1994, January 21). Don't throw that couch out. *Psychiatric News,* p. 3.

Miller, E. J., & Rice, A. K. (1967). *Systems of organization.* London: Tavistock.

R. B. K. (1994, February 4). NME drops out of hospital business. *Psychiatric News,* p. 9.

Satel, S. (1994, May). Apocalypse soon? *Psychiatric Times,* p. 19.

# Afterword

*Faith Gabelnick, Ph.D., Edward B. Klein, Ph.D., and Peter Herr, M.A.*

There is little doubt that the transformations discussed in this book can and do raise the level of anxiety in the workplace, whether that workplace be a university campus, a multinational corporation, a health care facility, or a family-owned business. Yet, as Peter Schwartz (1996) has written in *The Art of the Long View* and in a more recent article in *Wired* magazine (Schwartz & Leyden, 1997), a rigid view of the future that holds on to highly predictable, organizational structures, social roles, or class systems ignores the enormous possibilities presented in a more open, process-oriented Information Age. Schwartz and others provide planning scenarios for consultants, leaders, teachers, and students that envision change, bringing increased prosperity to many in a global society, and they envision technology assisting in the transformation of all aspects of our living and working together:

> We are watching the beginnings of a global economic boom on a scale never experienced before. We have entered a period of sustained growth that could eventually double the world's economy every dozen years and bring increasing prosperity for—quite literally—billions of people on the planet.... These two megatrends—fundamental technological change and a new ethos of openness—will transform our world into the beginnings of a global civilization, a new civilization of civilizations, that will blossom through the coming century. (Schwartz & Leyden, 1997, p. 116)

Their vision of a global society is quickly becoming a reality. At the edge of the 21st century, individuals are becoming more used to working in teams, not only in face-to-face teams, but in work groups connected technologically throughout the world. Consultants, therefore, are being asked to work in virtual environments where the production site may be very distant from the leadership and information site. Space, a traditional conceptual container for organizational dynamics, is already giving way to cyberspace. Thus, the challenge for consultants and leaders will be to understand the new space in which these dynamics play out and how leaders and followers are learning to cope and to prosper in environments that no longer "hold" us in any predictable way.

The nature of the consultant roles are therefore undergoing change. Like the organizations who hire consultants, this professional role requires extraordinary adaptability. Our authors work in different ways with CEOs, and future consultants will have to help leaders to work at the edge of their imagined competence in a world that has little patience for connecting reflection with action. Increasingly, the phrase "institutional transformation" is being used to focus on the *process* of continuous change rather than the end point. While consultants may be asked to work with organizations on plans with concrete outcomes, our authors report that the psychodynamic processes of the individuals who comprise the organization collectively create a dynamic system that does not submit easily to concrete controls or predictions. This focus on the organization as a living system offers great opportunity for a consultant to form a partnership with leadership and to work to guide change rather than mandate it.

Many of the chapters in this volume focus on this transformation. Therefore, more and more consultants find themselves in the roles of "coach" to a CEO or other executive. The consultant works on an ongoing basis to explore the development of the leader and the enactment of that development institutionally. Consultant/coaches may partner with leadership for many years, offering a private place for reflection and learning. This is an exciting and often profound partnership, but it has its hidden costs also. Leaders, consultants, and organizations are being pushed to the boundaries of their competence, experience, and imaginations. Not surprisingly, consultants themselves employ their own coaches, or "shadow consultants," to support their work. They take up Vaill's (1996) idea of "learning as a way of being" and seek out other colleagues to process their own feelings and insights as they navigate in complex organizational environments, working at the boundary of the person and the system. Consultants absorb the trauma and exhilaration of change.

Systems as we have known them are transforming into new networks, flexible and adaptive, and the psychological challenges for navigating in any of these roles in the new workplaces are still being discovered. For example, the emerging leaders, those in their 30s and early 40s, are inhabiting a world that is increasingly driven by technology, and *their* successors will find cyberspace, virtual work environments to be the norm. Rather than a heroic captain of industry leading change, young adults and, in some cases, children, are driving this transformation. The technology revolution was to a large degree spearheaded by teenagers. These teenagers are now nearing 50, and their entrepreneurial successors are still 2 or 3 decades younger. Children of 3 are learning to read and do mathematics using sophisticated technology, and their expectations of their parents, their schools, and eventually themselves as leaders will be very different from what we today can envision. Similarly, consultants are also challenged to work with CEOs who may be generationally distant from them or who may be leading organizations where their employees are already inhabiting a very different world.

One of the notable transformations in the workforce, therefore, is that individuals in their 50s are actively considering new career paths as the corporate environment dispenses with them or they with it. Given an increasingly longer life expectancy, these current baby boomers will be crafting at least one other career while paradoxically being considered "old" in their current work environment. Depending on their ages, consultants may find their identification valences being pulled in one way or the other, and therefore their view of the CEO similarly transformed. Overall, an enormous generational workforce transformation is underway, and consultants, too, will not be immune from the anxiety of being superseded by those who may have less experience than they and a very different worldview.

All of these transformations raise additional questions about the nature of community, social reform, and the new nature of leadership. We have argued in this text that leaders must be adaptive learners and lead from a place of connectedness. While the pains associated with the loss of familiar work patterns and the anxiety accompanying the vista of unformed or emerging work roles are upon us, we have a wonderful opportunity to acknowledge change and move forward. The psychological capacity to manage this change remains an unknown but exciting challenge. We now have more and more individuals who are becoming millionaires by age 35. They are adjusting their work patterns and life choices and creating very different kinds of work systems. They are also considering ways of distributing their wealth to favorite social/political/economic causes. How will *they* work with consultants? What kind of leadership will emerge from those who will view the world as a series of projects and networks?

We assert in this volume that leadership for the 21st century must continue to involve a complex, psychodynamic perspective regardless of organizational structures. Ironically, as organizations become more process-oriented, we will have to depend more upon each other and in that way create multiple communities for connection and vitality. This prospect paradoxically leads us to join others in envisioning leaders who will enjoy creating new communities and will join these communities in new ways. Perhaps the emerging role of the consultant will be as a partner in a visioning process that includes not only the workplace (wherever that may be) but also the continuing, unfolding connection of that workplace to other aspects of one's life. In this way a more integrated, whole systems perspective could be created, "a simpler way," as Margaret Wheatley and Myron Kellner-Rogers (1996) assert, where processes connect, in A. R. Ammons's (cited in Wheatley and Kellner-Rogers, 1996) vision, "not so much looking for the shape/as being available/to any shape that may be/summoning itself/through me/from the self not mine but ours" (p. 9).

## REFERENCES

Schwartz, P. (1996). *The art of the long view.* New York: Doubleday.

Schwartz, P., & Leyden, P. (1997, July). The long boom. *Wired, 5.07,* 116–173.

Vaill, P. B. (1996). *Leaving as a way of being.* San Francisco: Jossey-Bass.

Wheatley, M., & Kellner-Rogers, M. (1996). *A simpler way.* San Francisco: Bernett-Kochler.

# Name Index

Adam, B., 235, 236
Agazarian, Y., 270
Alderfer, C., 38, 47, 62
Alderfer, C. P., 39, 47, 150
Alonso, A., 252, 260–261, 265–266
Ammons, A. R., 299
Armstrong, D., 163
Ashbach, C., 205
Astrachan, B. M., xvi, 38, 207
Astrachan, J. H., xvi, 40, 41–42, 48, 207
Auer-Hunzinger, V., 237

Bain, A., 106
Barry, T., 237
Bayer, R., 244, 245
Beck, A., 261
Bennis, W. G., 148
Berenbeim, R. F., 42
Berg, D. N., 168, 229
Bernard, H., 259
Bhagwandas, R., 190
Bion, W. R., 58, 59, 64–65, 67, 68, 106, 115, 121–123, 126, 140, 148, 158, 164, 185, 194, 212, 241, 256, 270

Bond, J. T., 227
Bowers, P., 38
Bowles, M. L., 246
Brabender, V., 270
Brannigan, M., 225–226
Bridger, H., 100, 168
Budman, S., 259
Burlingame, G., 263

Caplan, G., 134
Caplan, R. B., 134
Carr, W., 115, 121, 124, 126
Carter, R. T., 69
Casemore, R., 168
Chabert, J.-L., 8
Chapman, J., 177, 190
Chatwin, M. E., 240
Cheevers, J., 113
Chemia, K., 8
Chisholm, R., 164
Cohen, C., 287
Cooke, R. A., 109–110
Cooper, R., 236

Covey, S., 228
Crystal, S., 235, 237

Dahan Dalmedico, A., 8
Dalgleish, J., xviii
Damasio, A., 7
Darrow, C. N., 280–281
Davidoff, D., xvii
Davis, M., 259
Demby, A., 259
D'Emilio, J., 61
Dies, R., 257
Duffe, J., 34
Dumas, R. G., 82–83
Dunham, H. W., 277
Dyos, G., 168

Eden, A., 168
Eisold, K., 77
Elden, M., 164
Ellis, A., 261
Emanuel, D., 170
Ergas, Y., 237
Esterson, A., 294
Evans, R. J., 247
Ezriel, H., 143

Fallon, A., 270
Fanon, F., 76
Farmer, P., 235–236, 247
Feldstein, M., 259
Fenichel, A., 68, 76
Fine, B. D., 119–120
Fineberg, H. V., 235, 245
Fitz, A., 38
Fleming, J. L., 68, 76
Foley, C., 190
Fouad, N., 69
Foucault, M., 6
Freud, S., 5, 6, 11, 16
Friedman, D. E., 227
Fuhrman, A., 263

Gabelnick, F., xiv, 206, 215

Gagnon, J. H., 245
Galinsky, E., 227
Gates, H. L. Jr., 66, 72
Gazda, G., 263
Gilkey, R., 134
Girard, R., 12n, 14
Gold, U., 206
Gould, L. J., xvii, 106, 138
Green, Z., 59, 67, 75, 76
Grinberg, L., 116, 123
Gutmann, D., xvi, 6, 15

Hale, A., 269
Harding, S., 140
Harris, M., 72
Harshbarger, D., 42
Hausman, W., xix
Herbert, J. I., 42
Herdt, G., 235, 245, 246
Herr, P., xiv, xviii–xix, 215
Hirschhorn, L., 124, 172–173
Hirschorn, L., 134
Hirschowitz, R. G., 275
Horwitz, L., 121–122
Huber, J., 237

Iglehart, J. K., 37
Imhof, A. E., 236

Jackson, M., 235, 237
Jullien, F., 10

Kaiser, G., 235
Kassvier, J. P., 37
Katzoff, J., 37
Kellner, K., 168
Kellner-Rogers, M., 299
Kerg, D. N., 228
Kernberg, O., 135, 143, 148, 149–150
Kets de Vries, M. F. R., 113–114, 127, 141, 149, 158
Khaleelee, O., 184–185
Khan, M. M. R., 243
Kiev, A., 277

Kindig, D. A., 37
King, A., 103
Klein, E. B., xiv, xviii, 38, 39, 138, 143, 150, 151, 154, 194, 215, 270, 280–281
Klein, M., 58, 108, 121, 194
Klein, R. H., xix, 251, 259, 270
Kleinman, A., 235–236, 247
Kohut, H., 147, 156
Kram, K. E., 229
Krantz, J., 212, 227, 228–229

Lacan, J., 25
Laing, R. D., 294
Langley, C., 190
Lawrence, W. G., 103, 106, 206, 243
Leigh, P., 37
Leszcz, M., 261
Levinson, 114
Levinson, D. J., 280–281
Levinson, H., 134, 136, 143, 278
Levinson, M. H., 280–281
Leyden, P., 295–296
Lieberman, M., 152, 156, 270
Lindenbaum, S., 244, 246
Long, S. D., xviii, 163, 164
Luske, B., 233

MacDonald, H., 52, 54
Malone, B., xvii
Manoussakis, G., 190
Mays, H. L., 37
McAuley, J., 168
McCallum, 257
McCollom Hampton, M., 229
McCollum, M., 114–115, 125
McGregor, D., 38
McGuire, M., 190
McInnes, G., 190
McIntosh, P., 61–62, 69
McIntyre, J., 280
McKee, B., 280–281
Mendel, G., 21
Menzies, I. E. P., 107, 134, 246

Miller, D., 113–114, 127, 141, 149, 158
Miller, E., 168, 184–185
Miller, E. J., 124, 205, 213–214, 229, 276
Miller, J., 38
Mirvis, P., 168
Moeller, M. L., 239, 244
Molesme, R., 140
Money-Kyrle, R., 127
Moore, B. E., 119–120
Moreno, J. L., 261–262
Morgan, G., 242
Morrison, T., 54, 60
Moss, S., 168

Nelson, B., 229
Newton, J., xviii, 163
Nicholas, M., xix, 262, 270
Noumair, D. A., xvi, 68, 76

Ormont, L., 259

Pierre, R., xvi
Piper, W., 257
Poupart, R., 245
Powell, M., 170

Racker, H., 119–120
Reed, B., 163, 237
Reed, G., xvi
Revens, R., 164
Rice, A. K., 58, 124, 194, 276, 277, 292
Rioch, M., 58
Rioch, M. J., 106
Rippinger, J., 139
Rivo, M. L., 37
Roberts, C., 190–191
Rosenberg, C. E., 34, 35, 36, 236–237, 246–247
Rosenfeld, H., 235
Rosenthal, J. S., xvii, 124–125
Rubin, G., 61
Rutan, S., 260–261

Sampson, E. E., 61, 63, 67

Samuelson, R. J., 41
Satel, S., 288
Say, J. B., 5
Sayles, J., 103
Schaar, J., 33
Scharff, D., 125
Scharff, J., 125
Schermer, V. L., 205
Schneider, B. E., 237
Schoeck, H., 176
Schumpeter, J. A., 5–6
Schwartz, H. S., 134
Schwartz, P., 295–296
Segal, H., 243
Seidl, O., 240
Shapiro, E., 115, 121, 124, 126
Shepherd, H. A., 148
Sherif, C. W., 60
Sherif, M., 60
Sievers, B., xix, 233–234, 236, 237
Singer, D., 259
Siporin, M., 263
Skolnick, M. R., 59, 67, 75, 76
Smith, A., xvii–xviii, 5
Smith, A. H., 136
Soldz, S., 259
Sontag, S., 234–235, 236, 245
Spicer, B., 170
Springer, T., 259
Stapley, L., 168
Starr, P., 37
Steiner, J., 243
Suarz, R., 228
Sullivan, H. S., 115, 261

Tabor, T. D., 109–110
Tate, D., 237
Thomas, R. R. Jr., 52
Toseland, R., 263
Tournier, M., 25–26
Tucker, L., 47
Tucker, R. C., 47
Turquet, P. M., 106

Vaill, P. B., 297

Waldvogel, B., 240
Walker, A., 70
Walsh, J. T., 109–110
Weber, M., 186
Weigand, W., 237
Weinel, E., 243
Wells, L., 58
Wheatley, M., 299
Whitaker, D., 152, 156, 270
Whitcomb, M. E., 37
White, J. B., 225–226
Williams, M., 68, 69, 71, 77
Willshire, L., 168
Winnicott, D. W., 229, 243
Winter, R., 164
Witkin, A., 245

Yalom, I. D., 204, 255, 259, 261
Young, D. R., 134

Zaleznik, A., 134
Ziegler, J., 236

# Subject Index

Acquisitions, individual responses to, 41–42
Action learning, 162, 164–166
  cycles of, 165
  group competition in, 182–183
Advisers, 28–29
African Americans
  versus Blacks, 54
  conflict with other minorities, 72–73
  disowned projections onto, 60
  diversity and, 52–53
Aggressor, identification with, 70–71
AIDS
  organizational impact of, 237–247
  societal impact of, 245–247
  time of, 233–237
  war against, 247
AIDS treatment organizations, xix, 233–247
American Group Psychotherapy Association, requirements of, 268
American male identity, 60

Antidepressants, 278
Antipsychotics, 278
Anxiety
  with change, xvii
  in downsizing organizations, 212–214
  in workplace, 295
Anxiolytics, 278
Appetite, 23
As if perspective, 106
Assassinations, 13
Assessment, 86–92
  results of, 88
Authoritative followership, 19
Authority, 194
  diversity and, 51–77
  fantasies surrounding, 173
  issues of, 182
  paranoid relationship with, 228–229
  sharing of, 20–21
  split, 186–187
Authority-follower relationship, idealized, 33
Authorization, meaningful, 68, 73–75

Autocratic leaders, 14–15
Autonomy
  change as predator of, 92
  local, 47

Barricades, 12
Barriers, 12
Basic assumption beliefs, 64–65, 194
Basic assumption theory, 105, 106, 205–206
Benedictines, traditions of, 139–140
Black box, 8–9, 11
Black consultant, client relationship of, 82–83
Black networks, 90
  power of, 91
Black plague, 235
Boards, 34
Bottom-line leaders, 46
Boundaries, 194, 203
  collaboration across, 57
  confusion of, 19
  in crisis, 22–23
  external and internal, 195
  protection of, 43
  shifting, 148–149
Boundary management, 124
  in consultant-client relationship, 130
Business
  organizational processes in, 133–134
  restructuring in, 134–136
Businesslike behavior, 179

Capitation systems, 286
Career protection, xvi
Caring roles, 94–95
Case management, 165
Catholic religious orders, consulting in, xvii–xviii, 133–157
Change
  anxiety over, xv, 295
  collaborative management in, xix–xx
  continuous flow of, 94
  in health care system, 211
  information technology in, 188–189
  with intact organization purpose, 179
  integration of, 94
  in leadership, 203–204
  in psychiatric professional marketplace, 275–294
  in religious orders, 139–140
  resistance to, 67–75, 170
  rhetoric versus reality of, 91–92
  understanding and managing, xiv–xv
Chaos theory, 8
Charism, 138–140
Charismatic leadership, 147–151
Cistercian order, 140
Civil rights movement, 62
  minority groups in, 72
  protests of, 23
Clinical care coordinators, 283–284
Cloning process, 3–4, 30
Club Mediterranée, 29–30
Coaching, 297
Coal industry, nature of and psychic manifestations, 107–108
COALCO organization
  death of, 109–110
  group dynamics of senior team of, 105–109
  intervention for, 101–104
  management team dynamics in, 104–105
  past and present, 98–99
  planning intervention for, 99–101
Cognitive-behavioral groups, 260, 261
Collaboration
  complexity of, 185–186
  model of, 67, 75–76
  understanding of, 186–189
Collaborative action research project, 166–167

in changing government
    organization, xviii
  providing holding environment
    for, 168–172
Collaborative environment, 162–163
Collaborative management styles,
    xix–xx
Collaborative state of mind, 185–189
Collective state of mind, 164
Commensal relations, 185, 186
Commitment, to communities, 47–48
Communication
  continuity of, 44
  missed, 89
  value of, 227
Community, dynamism of, 138–140
Community hospitals, nonprofit,
    289–290, 293
Compensation
  executive, 45–46
  structure of, 107–109
Completion bonus dilemma, 107–109
Complex systems, 194
Complexity, institutional, xiii–xiv
Concordant identification, 120
Conflict, 221–223
  in group therapy, 256
  in management, 217–218
  nurtured, 90–91
  reframing of, 12
  underlying diversity, 52–53
Conformation, 30
Connectedness, xx
  in group therapy, 259–260
Consensus leadership, 150
Constituency building, 62
Consultant. *See also* Shadow consultants
  accidental, 211–230
  changing role of, 296–297
  client's perception of, 125–126
  as container, 193–207
  ego boundary permeability of,
    122–123
  family business owner and, 117–120

  as messiah or savior, 117–118
  as Other, 25–26
  as participant observer, 115
  partnership with client, 115
  permeable ego boundaries of,
    115–116
  power of, 206
  projecting onto, xviii
  role of, xiv–xv
  in sustaining triangle, 5–13
  unresolved developmental issues
    of, 127–128
  use of self by, 116–124
Consultant-client relationship
  countertransference in, 120
  in family business, 117–120
  levels of, 130
  projective identification in, 121–122
  transference in, 119–120
Consultation
  assessment results in, 88
  conflict in, 221–223
  developmental stages in, 194–202
  diversity, 54
  during downsizing, 218–220
  failure of, 94–96
  first task in, 4
  focal conflict model of, 152–153
  influences on outcome of, 114–115
  inner dialogue technique in,
    113–131
  in institutional transformation,
    3–30, 24–30
  issues of, 88–92
  methodology of, 87
  multiracial, 81–96
  of nonconsultant, 224–225
  projective counteridentification in,
    123–124
  reality checks with client in, 130
  regression in service of, 121, 123
  request for, 83–86

scapegoating in, 220–221
stated charge of, 86
termination process in, 202–203, 204–205
time boundaries of, 193, 205
during transition, 216–218
unstated charge of, 86–87
Consultation team, multiracial, xvii
Container, xx, 59, 115
consultant as, 193–207
failure of, 206–207
of self-esteem, 147
for staff emotions, 225
Containment, 23, 115, 168–172, 184
Content task, 100
Continuity, communications, 44
Contract negotiation, 194–196, 203
Corporations, psychological contract in, 37–40
Corrective emotional experience, 261
Cost-containing mental health care organizations, xix
Counter-Reformation, 139–140
Countertransference, xviii, 116, 125, 127, 206
in consultant-client relationship, 120
in family business consultation, 119–120
in long-term consultation, 193
Covert process, interpreting, 194, 197–200, 203
Creative arts therapies, 260, 262
warm-up techniques of, 269
Creative destruction process, 5
Crisis, leadership in, 22–23
Crowd theory, 16
Cultural diversity, xv, 53
Culture
AIDS impact on, 246–247
organizational, 175–181
Customer service, importance of, 225–227

Day treatment settings, 269–270

Death
anxiety over, 243
awareness of, 240–241
coping with, in HIV group home, 241
desire and, 15–16
fear of, 235–237
Decision making
emotions in, 7
by group, 135
joint, 38
transferential patterns in, 113–114
Defenses, 58–59
against guilt of coal mining industry, 107
in psychodynamic theory, 267–268
Deinstitutionalization policies, 287
Delta Air Lines, downsizing problems of, 225–226
Democratic management, 38–39
Denial, 107
Dependability relationship, 23
Dependency, 202–203
basic assumption, 106, 107, 256
culture of, 241
defensive reaction against, 155–156
on leadership, 147–148, 205–206, 212–214
on omnipotent leaders, 150–151
of religious communities, 141–142
staff-patient, 198–199
Dependency-based transference, 146
Depressive structure, 11
Depressive-impulsive style, 145–147
Desire, 15–16
Development
individual, 265–266
normal, 264
stages of, in consultation process, 194–202
unresolved issues of, 127–128
Didactic training, group therapy, 267–268
Difference, managing, 90–91

*The Dilemma of Being Both Victim and Victimizer* (Williams), 68
Direction, need for, 88
Discernment, 18–19
Discovery task, 161–162
Discrimination
  institutional, 57
  against people with AIDS, 235–237
Disempowerment, 51
Disengagement, 105
Disowned projections, 60
Distrust, pervasive, 89
Diversity. *See also* Minority groups
  conferences on, 55–56
  conflicts underlying, 52–53
  group relations work and, 76–77
  issues of, 53–54
  myth of redress and, 64–67
  new model of, 75–76
  perspective on, 56–57
  politics of identity and, 51–77
  resistance to change and, 67–75
  resource allocation and, 62–63
  task forces for, 74
  theoretical orientation to, 57–59
  training in, xvi
  verifying commitment to, 86–87
Diversity and Authority Conference, Teachers College, 56–57
Divestiture
  coping with, 97–111
  managing, xvii
Downsizing, 216, 218–220
  consultation for, 211–230
  customer satisfaction and, 225–227
  in health care organizations, xviii–xix
  scapegoating with, 220–221

Egalitarianism, 176–177
Ego boundaries, permeability of, 115–116, 122–123
Ego ideal, 16
Ego mirror, 25

Emotions
  in decision process, 7
  of family business members, 115
  in family business relationships, 118–119
Empathy, in consultation for family business, 122–123
Employee dissatisfaction, 224–225
  causes of, 227–228
  in downsizing organizations, 212–214
Employees
  attitude survey of, 221–224
  new way of dealing with, 228–230
  as source of innovation, 228
  transfers of, 219
  turnover of, with downsizing, 220–221
Empowerment, 51–77
  need for skills in, 85
Energy company, divestiture of, xvii
Ethics, vision and, 17–18
Ethnic diversity, 53
Ethnic groups, xv, 53–54
  in university setting, xvi
European Union, 18
Executive compensation, appropriate, 45–46
Executive success, bottom line and, 225–227
Experience
  irrational and unconscious aspects of, 165–166
  learning from, 164–166
External boundary issues, 201–202
Externalization, 59

Facts, discovering of, 11
Failure, unresolved grieving over, 110–111
Family business, xvii
  boundary management in, 124
  influences on consultation outcome for, 114–115

inner dialogue in consultation for, 113–131
narcissistic extension of founder in, 128
patriarchal leadership of, 117–119, 130–131
response of, to stressors, 114
transference patterns in, 114
Favoritism, 90
Feedback process, 179–180
Fee-for-service payments, 286
Fiat, Agnelli family in, 22
Fight/flight basic assumption, 106, 107
Focal conflict, 156
in case example, 153–157
model of, 152–153
Followers, capacity to have, 19
Followership, 228
For-profit health care organizations, 283–285, 286–287, 288–289
Founder
connection to ideals of, 136–138
group assumptions and leadership style effects of, 140–143
idealization of, 142
ideals of, versus society, 137, 138–140
legacy of, 133–159
organizational restructuring and, 134–136
passing on ideals of, 157–159
Fragmentation, 68
among subgroups, 71–73
defense against, 243–244
Franciscans, 139
Fusion, desire for, 238–239

Gay rights, 53
Gay/lesbian community, 65–66
in civil rights movement, 72
disowned projections onto, 60
Gender

issues of, 44
relations of, 202–203
Gender differences, xv
Generative leadership, 16–24
Germany, AIDS treatment organizations in, xix
Global society, 295–296
Goals, long-term, 45
Good enough mothering, 194, 229
Government organization, change in, xviii
Grass roots advocacy, 74–75
Great Britain, Thatcher's leadership in, 15
Grief, unresolved, 110–111
Group. *See also* Minority groups; Support groups
abandonment of defenses of, 68
behavior of, in business and religious communities, 133–134
boundaries of, 203
clarity of task and role in, 163–164
cohesion of, 136–137, 259
culture versus founder's ideals in, 137–138
development of, 270
diversity and relations in, 76–77
dynamics of, in management team, 104–105
founder's ideals and assumptions of, 140–143
founder's legacy passed on through, 157–158
here-and-now tensions of, 152
leadership of, xix
overt and covert processes of, 203
self-appraisal of, 104–105
splitting in, 196–197
Group process
depressive-impulsive styles of, 145–147
narcissistic styles of, 147–151
obsessive-paranoid styles of,

144–145
  regressive leadership in, 143–151
Group relations model, 57–59
Group therapists
  obstacles to training, 253–255
  scarcity of, 252–256
  training for 21st century, 251–272
Group therapy
  applications of theory, 270–271
  benefits and disadvantages of, 252
  definition and value of, 257
  dependence on leader in, 256
  economic and time constraints on, 271–272
  economics of, 251–252
  effects of scarcity of trained therapists on, 255–256
  essentials of training in, 264–271
  five key dimensions of, 256–258
  future for, 251–252
  group composition in, 257
  helping members do their work in, 268–270
  history of, 271–272
  incompetence in, 254
  intermember support in, 256
  length of treatment in, 258
  modality and setting of, 257
  need for research on, 262–264
  nontheory-based, 258–260
  problem addressed in, 257–258
  supervision and didactic training for, 267–268
  systems approach in, 270
  in teaching hospitals, 254–255
  theory-based, 260–264
  therapeutic goal of, 259
  types of, 258–260
  warm-up techniques for, 269
Group-as-a-whole, 59, 164
Group-environment relationship, 194, 200–202

Health care industry
  consulting in, xviii–xix
  downsizing in, 211–230
  psychological contract in, 34–37
Health care organizations. *See also* Health maintenance organizations; Hospitals; Managed care; Mental health organizations
  groups in, 34
  long-term consultation for, 193–207
  restructuring of, 37
  staff of, 34–35
Health maintenance organizations, 286–287
  changes in, 293
  changes required by, 211
Health settings consultation, xvi
Heavy hitters, recruitment of, 45
Here-and-now activation, 261
Hermeneutics, 11
Heuristics process, 11–12
Hidden elements, 12
HIV group home
  awareness of mortality in, 240–241
  dependency culture in, 241
  failure of management in, 242–243
  psychologist's role in, 238–239
  psychosocial dynamics of, 239–237
  responsibility in, 244–245
  splitting of staff in, 240–242
HMOs. *See* Health maintenance organizations
Holding environment, 168–172, 184
  in group therapy, 259–260
Homeostasy principle, 7
Homophobia, 60
Homosexuality, identity and, 61
Hospitals
  administration, changes in, 283–284
  changes in, 279–280
  downsizing in, 218–220
  for-profit, 283–385
  nonprofit, 285–286, 293
  overworked staff of, 285

professionalized system of, 35–36
reformers of, 35
social system dynamics of, 215–216
state, 287, 292–293
teaching, 288
transition in, 216–218
VA, 288
work relations in, 36
Human behavior, fantasy of monitoring or forecasting, 6–7
Humanistic paternalism, 39–40
Hysteria, 145. *See also* Depressive-impulsive style

Idealization, 142, 234
Identification
  with aggressor, 70–71
  concordant, 120
Identity
  of oppressed groups, 60–61
  organizations and, 214
  political use of, 61–64
  politics of, 51–77
  social dynamics of, 59–61
Immortality, illusion of, 234
Inconsistency, 89–90
Individual development, 264
  in group therapy, 265–266
Inequality, 196
Infant-mother relationship, 194
Information technology, 188–189
Inner dialogue
  for family business consultation, 113–131
  strategies to promote, 130
Innovation, 21–22
Institutions
  enforcing of boundaries of, 19
  loyalty to, xix
  need for transformation of, 5–13
  pathological, 243–247
  transformation model of, 4–30, 296–297
Instrumental reasoning, 186

Intergroup tensions, reduced, 44
Internalization, 59–60
International Forum for Social Innovation, Institutional Transformation working conference of, 27
Interpersonal/interactional group therapy, 260, 261
Interpretative stance, 115–116
  negotiated, 124–125
Intervention
  in divestiture process, 101–104
  planning of, 99–101
Interviews, senior management, 101–102
Intimacy, in consultation for family business, 122–123
Irrational material, in transformation, 7–9

Jesuits, 139–140
Jim Crow legacy, 91
Job security, xvi, 42–43
  loss of, 48
Joint Commission on Accreditation of Health Care Organizations survey, 222–223

"L.A. Four," 52
Languages, xv
Latino studies, 54
Layoffs, dealing with, 48
Leader-follower relationship
  evolving, 33–49
  idealized, 33
Leaders. *See also* Leadership
  assassination of, 13
  in crisis, 22–23
  definition of, 4
  generating, 21
  Januslike, 194
  larger-than-life characteristics of, 194
  loss of, 202–203

self-destruction of, 14
visionary, 16–17
Leadership. See also Management; Senior management
  advisers in, 4, 28–29
  changing, 203–204, 298–299
  characteristics of, 17–24
  collaborative style of, xix–xx
  dependency on, 147–148, 202–203, 205–206
  distortions of, 14–15
  effective, xix
  egalitarian role of, 145–146
  ethics of, 11
  etymology of, 4
  failed, 29–30
  failure projected onto, 107
  founder's legacy passed on through, 157
  generative and generous, xvi, 3–30
  lack of, 84–85
  open-ended facilitative, xv
  organization and, 212–214
  psychiatric practice changes and, 289–291
  regressive, 143–151
  sociotechnical and psychological changes in, xiv
  for 21st century, 109–110
  temporary, xix
  transformation and, 13–16
  transition of, 216–218
  uncertain, 200–202
Leadership styles
  founder's ideals and, 140–143
  outdated, 228–229
  transferential patterns in, 113–114
Learning, action and, 162, 164–166
Learning-from-experience, 187
Levinson Institute, 277–278
Local autonomy, 47
*Locum tenens,* xix
Long-term consultation, countertransference in, 193

Long-term goals, 45
Los Angeles
  riots of, 23
  Unified School District of, 63
Loyalty, 90, 230
  end of, xiv
  loss of, xix, xvi

"Mammy" role, 82–83, 87, 94
Managed care, 284–285
  group therapy in, 263–264
  insurance programs of, 290
Management. See also Leadership; Senior management; Shared management
  conflict in, 217–218
  effectiveness analysis of, 85
  flatter structures of, 188
  in HIV group home, 242–243
  modern concept of, 6
  in reorganization and downsizing, 213–214
  shared, 19–20, 29
  theory Y, 38–39, 39
Management team, dynamics and functioning of, 104–105
Management theory, 38
Management-employee relationship, tension in, 44
Managers, health care, 34
*Matewan,* 103
Medicaid fees, 284
Medical center, multiracial consultation for, 83–96
Medical therapies, 278, 279–280
Medicare fees, 284
Mental health organizations
  administrative staff of, 283–284
  changes in, 282–283
  economic and time constraints on, 271–272
  market changes in, 291–292
Merger fantasy, 147
Mergers, xvii

individual responses to, 41–42
Meritocracy, myth of, 61–62, 69
Microsoft, sustaining triangle in, 24–25
Middle managers, elimination of, 213
Minority groups. *See also* African-Americans; Diversity; Oppressed groups
　conflict among, 71–73
　definition of, 53–54
　failure to incorporate, 44, 47
　lack of meaningful authorization in, 73–75
　oppression of, 63–64, 69–70
　victimizer role in, 70
Minority identity, 63
Miscommunication, 89
Mortal-immortal split, 234, 245
Mortality
　awareness of, 240–241
　denial of, 236–237
Moses, leadership of, 19
Mother-infant relationship, 194
Mourning
　at loss of organization, 110–111
　unresolved, 110–111
Multiculturalism, 53
Multiracial consultancy teams, xvii, 81–96

Narcissistic group process, 147–151
Narcissistic injury, redress myth and, 64–65
Narcissus, 235
National Alliance for the Mentally Ill, 287
Networks, 229
　segmented, 89
Nightingale, Florence, professionalism of, 35
Not me, 60
Nurses
　downsizing staff of, 218–220
　high turnover rates of, 220–221
　overworked, 285
　professionalization of, 35–36

Object relations approach, 205
Objectives
　as fantasy, 7–8
　short-term, 47–48
Obsessive-paranoid style, 144–145
Open systems theory, xv, xx, 7
Open-ended contract, 195–196
Oppressed groups
　fear of change in, 67–68
　identity of, 60–61
Oppressor-oppressed polarization, 76
Organizational consultation, xvi
　frame of, 151–152
　research consultation and, xviii
Organizational restructuring, 40–42
　Vatican II impact on, 134–136
Organization-as-a-whole, 164, 166
Organization-in-experience, 163, 165–166, 189
　reestablishing, 178
Organization-in-the-mind, 163
Organizations
　AIDS impact on, 237–247
　boundary changes of, xviii
　culture of reflected in steering committee, 175–181
　dynamics of, 215–216
　emotional health of, 212
　founder's legacy and behavior of, 133–159
　identity through, 214
　leadership of, 212–214
　pathological, 243–247
　psychic effects of loss of, 109–111
　transformation of, 4–30
Other, 59–60
　AIDS and, 235
　consultant as, 25–26
　presence of, 161–190
　working in presence of, 186–190
Outreach offices, 284–285

Outsiders, 21
Overt process, interpreting, 194, 197–200, 203

Pairing basic assumption, 106, 107
Paranoid, solidarity of, 244
Parasitic relations, 185
Participant observer, 115
Participation, 230
Participatory action research project, 161–190
Participatory action research working laboratory, 167
Path, maintaining, 23–24
Patient management systems, 279
Payment systems, 286
Persecutor, 14–15
Personal history, xvii
Personality model, 267–268
Physicians, 34
Plan of action, 93–94
*Playing in the Dark* (Morrison), 60
Pleasure principle, beyond, 5–6
*Possessing the Secret of Joy* (Walker), 70
Postassessment phase, 93–94
Power, 61
  cultural assumptions about, 61–62
  distribution of, 75–77
  scarcity assumption and, 63
Power relations model, 75–76
Powerlessness, feelings of, 228–229
Praxis International, Institutional Transformation working conference of, 27
Preassessment phase, 83–86
Preferred provider organizations, 211
Presumed successors, 20–21
Primary task concept, 276
Privilege, denial of, 69–70
Process task, 100
Professional marketplace, changes in, 275–294
Profitability, 292
Projection, in self-concepts, 59–60

Projective counteridentification, 116, 123–124, 128–129
Projective identification, 58–59, 116, 121–122, 129, 194
  in HIV group home staff, 242
Projective processes, 138
Psychiatric disability, 177
Psychiatric institutions, change in, 279–280, 293–294
Psychiatric practice, changes in, 280, 282–294
Psychiatrist, locum tenens, 276, 281–294
Psychiatry, changes in, 278–281
Psychiatry departments, quality of group therapy in, 254–255
Psychic exhaustion, 13–14
Psychoactive drugs, 279
Psychoanalytic psychotherapies, 260–261
Psychodrama, 260, 261–262
  warm-up techniques of, 269
Psychodynamic culture change, xv
Psychodynamic theory, xx
Psychodynamic therapy
  group, 260–261
  questioning value of, 279
  training in, 267–268
Psychoeducational groups, 260
Psychological change, leadership, xiv
Psychological contract, xvi
  changing, 33–49
  in corporations, 37–40
  in health care industry, 34–37
  shared silent, 40–42
  trends in, 46–49
  trust building in, 42–46
Psychologist, roles of, with HIV-infected people, 238–239
Psychopathology
  in group therapy, 264, 265–266
  level of chronicity of, 257–258
Psychosocial dynamics, in HIV group home, 239–237

Psychotherapy, individual, 265–266

Race issues, 44
  as coincidental, 85–86
  in consulting, 81–96
Racial relations, 202–203
  tensions in, 52–53
Racism, 47
Rational-Emotive Therapy, 260, 261
Reactive fear, 154–156
Reality principle, 106
  losing parameters for, 244
Reasoning, instrumental versus value, 186
Redress, myth of, 64–67
Reductionism, 63–64
Reference group, 176–177, 178
Regression
  with downsizing, 220
  in leadership, 143–151
  in service of consultation, 121, 123
  of staff, 228–229
  unwillingness to accept, 11
Regression-prone leaders, 149–150
Relatedness, types of, 185–186
Religious communities
  businesses and, 133–134
  consulting in, xvii–xviii
  focal conflict in, 153–157
  group assumptions and leadership styles in, 140–143
  here-and-now group tensions in, 152–153
  regressive leadership styles in, 143–151
Repetition, 5–6
  in succession, 22
Representations, 7
Research, 166–167
  collaborative, 162
  consultation on, xviii
  stages of, 167
  on theory-based group therapy, 262–264

Resistance, to transformation, 8
Resistant people, 170
Resource allocation, diversity and, 62–63
Responsibility
  in HIV group home, 244–245
  stress on, 261
Retreat
  in divestiture process, 102–104
  planning for, 99–100
  purpose of, 100
A. K. Rice Institute, 277–278
Role(s)
  AIDS impact on, 237–247
  boundaries of, 22–23, 173
  clarity of, 162, 163–164
  continual negotiation of, 173–174
  negotiation of, 172–173
  overload of, 220
  retreat from boundaries of, 173
  task and social, 194
  tension between, 174–175
Roman Catholic church
  charism of, 138–140
  commitment to founder's ideals in, 136–138
  emotional upheaval in, 135
  Vatican II impact on, 134–136
Rumors, 218

Salt-and-pepper consulting team, xvii, 81–96
Savior, 14
Scapegoating, 14, 213–214, 220–221, 228
  addressing causes of, 46
  by narcissistic leader, 148, 149–150
  of subgroups, 150
Scarcity, assumption of, 61–62, 63
  victimization and, 65
Schizoid detachment, 107
Segregation tradition, 91
Self-awareness, xvii
Self-destructive behavior, 14, 197

Self-disclosure, in group therapy, 259
Self-identities, 59–60
Self-interest, culture of, 235
Self-possession, myth of, 6
Self-selected groups, 182
Senior management. *See also* Leadership; Management
　in divestiture process, 100
　group dynamics in team of, 107
　interviews of, 101–102
　separation of, from staff, 227
Sexism, 47
Sexuality, AIDS and, 234–235
Shadow consultants, xviii, 130, 204, 297
Shared history, 43–44
Shared management, xvi
　generative leadership and, 3–30
Shared psychological contract, 39, 40–42
Short-term objectives, focus on, 47–48
Silicon Valley, history of, 228
O. J. Simpson verdict, 52
Social defenses, 194
Social dynamics, 59–61, 215–216
Social power, distribution of, 75–77
Social psychiatry, 278
Social reality, power to define, 61
Social roles, 194
Society, work world and, 48–49
Sociotechnical changes, xiv
Splitting, 58–59, 107, 194, 196, 234
　of Black female, 82–83
　in HIV group home staff, 240–242, 245
Stability
　need for, 88
　threaten of, 5–6
Staff
　regression of, 228–229
　roles of, 27
　shortages of, 219
　turnover of, 9, 216–217, 220–221
State of mind

　collaborative, 185–189
　collective, 164
　psychodynamics of, 162
Steering committee
　dynamics of, 176
　extending process work of, 181–185
　as holding environment, 168–172
　neutrality of, 170–171
　reflecting organization culture, 175–181
　as representative group, 176–178
Stereotypes, 194
Stewardship, 35
Stress, of changing organizational boundaries, xviii
Strikes, 199
　tensions after, 48
Subgroups
　competition among, 66
　fragmentation and competition among, 68, 71–73
Succession
　difficulties in, 21–22
　strategies for, 20–21
Superego, 16
Superleader, 14
Supervision, group therapy, 267–268
Support groups, 258–259
Sustaining triangle, 5–13, 24–30
　evolving roles in, 27–28
Swinburne University research project, 161–190
Symbiotic relations, 185
System
　diagnosis of, 203
　entering and diagnosing, 194, 196–197
Systems theory
　in group therapy, 264, 270

Task roles, 194
Task teams, levels of, 194
Tasks
　clarity of, 162, 163–164

negotiation of, 172–173
Tavistock concepts, 194
Tavistock Institute for Human Relations, group relations model of, 57–77
Team commitment, 47–48
Technological change, 295–296
Tension, intergroup, 44
Termination process, 202–203, 204–205
Thatcher, Margaret, leadership of, 15
Theme-centered group dynamic theory, 270
Theory Y management, 38, 39
Theory-based group therapy, 258–264
Therapeutic neutrality, 129
Therapeutic space, 129
Therapists
  patient relationships with, xviii
  resistance of, to group therapy training, 253–255
Three-party focus, 179
Time boundaries, 205
Training
  of group therapists, 251–272
  for group therapy, 266–267
  obstacles to, 253–255
Transference, 116, 174–175
  dependency-based, 146
  distortion of, 130
  in family business consultation, 119–120
  in leadership styles, 113–114
Transformation
  barriers to, 12
  blurred phases of, 8–9
  chaotic, 8
  definition of, 4
  desire of, 15–16
  leadership and, 3–30, 13–16
  objectives in, 7–8
  path of, 9
  peaks of, 15
  between shared management and generative leadership, xvi
  to sustain institution, 5–13
Trans-formation, 8, 10
Transformative leader, 13
Transition period, 216–218
Trust, 196, 202–203
  building, 42–46
  in family business relationships, 115
  in group therapy, 259–260
Turnover rate, 9, 216–217, 220–221

Uncertainty, 89–90
United States, transformation peaks in, 15
University, diversity training in, xvi
Unknown, acceptance of, 9–10

Value reasoning, 186
Values
  of diverse workforce, 47
  shared, 39–41, 43–44
Vatican II, 141
  impact on organizational restructuring, 134–136
Veteran's Administration hospitals, 288
Victim, 14
Victimization, 60–61
  medals of, 66
  redress myth and, 64–65
Victimizer role, denial of, 68, 69–71
Vision, 17–18
  discernment and, 18–19
  of founder, 157–158
  need for, 88
Vision statement, 93
Visionaries, 16–17
Voluntary institutions, dynamics of, 240–241

Wage pressures, 230
*Who's in Charge?*, 103
Women, disowned projections onto, 60

Work contract, changing, xiv
Work team, container function of, xx
Work (W) groups, 106, 194
Workforce. *See also* Employees; Staff
  involvement of, 93
  transformations in, 297–298
Working laboratory, 181–185
Working pair, 19–20
Working relations
  complexity of, 185–186
  conditions of, 163–166
  maturity in, 172–175
  maturity of, 162
Workplace
  changing psychological contract in, 33–49
  future of, 298–299

# About the Editors

**Edward B. Klein** is a Professor of Psychology at the University of Cincinnati; Faculty Member, Cincinnati Psychoanalytic Institute, and formally, Visiting Professor, Yale School of Management. He is coauthor of *The Seasons of a Man's Life*.

**Faith Gabelnick** is President of Pacific University in Forest Grove, Oregon, and President, A.K. Rice Institute.

**Peter Herr** is a therapist who is completing his Ph.D. in Clinical Psychology at the University of Cincinnati, Ohio.